A serious student's guide to
ALGEBRAIC EXPRESSIONS

Maksim Sokolov

MAXERGON
POWERED BY CENTURIES

Maksim Sokolov
Professor of Mathematics
Seneca Polytechnic
Toronto, Canada

ISBN: 978-1-7380649-0-8.

Typeset by the author in LaTeX.

Cover design: Maksim Sokolov. The drawing on the cover illustrates the cubic expansion:

$$(a + b)^3 = a^3 + 3a^2b + 3ab^2 + b^3$$

The drawing is by Drini.

To Oksana, Melania and Polina.

Without your support and patience, this book would not exist.

Without this book, your lives would have been a bit easier.

Contents

Introduction

The title of this book accurately reflects both its content and target audience. Specifically, this book:

- Focuses on the transformation, simplification, and evaluation of algebraic expressions. This book does not encompass the entirety of elementary algebra, however. For instance, it does not include topics such as equations and inequalities, and it does not involve complex numbers. In short, this book is exclusively devoted to building the core skill of manipulating algebraic expressions, which is foundational for any other mathematical study.

- Is a comprehensive guide that contains: (1) Detailed explanation of topics (2) Worked examples (3) Abundant exercises of varying difficulty levels (4) Solutions to all exercises.

- Is tailored for dedicated students such as yourself, who are ready to navigate the content diligently, embracing "productive struggle" with a positive attitude.

The book is appropriate not only for students starting their elementary algebra journey but also for those who have previously completed an elementary algebra course and are aiming to refine and amplify their proficiency with algebraic expressions. Particularly, students preparing for post-secondary education will discover this book to be especially beneficial.

Algebraic expressions are the foundational keystones for any quantitative subject. No matter what courses you wish to take later – let them be linear algebra, calculus, statistics, financial mathematics, or computer science – proficiency in these subjects will hinge on your ability to adeptly manipulate algebraic expressions. Thus, regardless of your chosen quantitative path, competence in manipulating algebraic expressions is a prerequisite.

Regrettably, there exists a substantial disparity between the content students are taught in secondary schools and the real-world challenges they encounter in colleges, universities, and workplaces. During their time in secondary school, students learn to manipulate only very simple algebraic expressions. The school curriculum does not pay enough attention to developing the necessary depth in this aspect. Yet, as students progress into their post-secondary studies, they are expected to handle much more complex expressions than they used to encounter in their secondary school. Many students of colleges and universities cannot keep up with the amount and speed of what they are taught, simply because they cannot transform algebraic expressions. What makes matters worse is that post-secondary institutions completely ignore this problem: professors do not have time to review algebraic expressions and teach under the assumption that their students are experts in this regard. Reaching the level of mastery expected from students by colleges and universities, or dictated by the realities of modern quantitative occupations, requires special preparation – and this is precisely where this guide can offer assistance.

You will begin by learning the foundational principles and progress towards achieving a genuine mastery of manipulating algebraic expressions. Through active engagement with this guide, you will establish a robust foundation to excel in any quantitative endeavor, whether it pertains to your academic pursuits or professional career.

To use this book, two prerequisites are essential. The first is an academic one: a solid grounding in arithmetic. The second prerequisite is to have an ample supply of paper and pens, as you will be working through a multitude of exercises. Indeed, as *serious* students know, attaining mastery in mathematics is less about reading and more about active involvement with exercises. You won't require anything else, not even a calculator.

Your journey

The whole book – every chapter and every exercise – has been carefully designed to guide you from start to finish: each exercise builds upon the previous one, gradually increasing in difficulty as your skills develop.

Chapter 1: You will begin by learning how to evaluate expressions of varying complexity. Here, no prior knowledge of formulas or techniques for transforming expressions will be required. Your arithmetic skills will be put to use and reinforced.

Chapter 2: You will delve into working with expressions that involve both whole and fractional exponents. This chapter will also introduce you to radicals. By the end of the chapter, you will confidently handle exponent operations of varying degrees of difficulty.

Chapter 3: The primary techniques for dealing with algebraic expressions, such as using the distributive property and collecting like terms, will be mastered here. Due to their paramount importance, you will engage in extensive practice to firmly grasp these techniques. In this chapter, you will also learn how to work with addition and subtraction of algebraic fractions.

Chapter 4: You will learn how to use crucial expansion equalities. These equalities play a pivotal role in expression manipulation, and you will gain extensive practice in applying them. You will engage with various techniques for expansion and factoring of algebraic expressions. By this chapter's conclusion, you will have become an expert in simplifying algebraic expressions of diverse complexities.

Chapter 5: This chapter focuses on the manipulation of polynomials. You will acquire proficiency in factoring and dividing polynomials, will learn about polynomial composition, and will gain insight into interesting concepts such as non-commutativity and one-sided distributivity.

Chapter 6: Expressions involving factorial, summation (sigma), and product notations will be your focus. These types of expressions are ubiquitous in almost all quantitative fields. As you engage with the exercises in this chapter, you will discover concepts such as arithmetic and geometric progressions, arithmetic and geometric means, variance and standard deviation, Euclidean distance, permutations and combinations, and the binomial theorem.

Upon completing all six chapters, you will have gained the proficiency required to confidently manipulate algebraic expressions at the level expected of students entering college or university.

Your approach to learning

You are a *serious* student, exemplifying the traits of a student who:

- Aspires to achieve *genuine* mastery of the subject matter.

- Recognizes that gaining *true* knowledge requires dedicated effort and hard work.

- Works through the material thoroughly, without skipping any information.

- Values the significance of working through exercises, both for gaining a deeper understanding and for developing agility.

- Maintains a positive attitude, being willing to undergo "productive struggle" and embracing the discomfort of confusion as a stepping stone towards growth.

- Understands that any academic setback is a positive sign that indicates that there is something new to learn.

- Works on errors systematically – never looks into solutions prematurely.

- Refrains from making excuses or resorting to statements such as, "I've never been good at math, so what can you expect from me?"

- Knows that recovering from a setback must be based on a well-defined *action* plan: "Based on the careful analysis of the result, going forward I will do X, Y, Z".

Your strategy

I recommend that you adhere to a well-established study strategy that has stood the test of time and is proven to lead to success. Specifically:

- In each chapter, read the section titled "Explanation" carefully.

- Analyze the section titled "Examples of worked exercises", verifying the process of the shown solutions carefully on paper.

- Engage in deliberate practice by working through exercises. Verify each answer you obtain using the answer key located at each chapter's end. Should your answer be incorrect, reattempt the exercise, taking care to revisit the corresponding material. In the event of repeated errors, consult the solutions presented in the latter portion of the book. Refer to the suggested solutions even if your answers are correct, as this might give you an insight into alternative approaches to solving the exercises. Completing all exercises is of the utmost importance, even if some of them appear monotonous. Mastery of algebra hinges on cultivating automaticity and developing "pattern recognition" skills that are obtained only through persistent engagement with exercises.

Should you observe any errors, typos, or wish to offer suggestions or feedback, kindly use the link **bit.ly/algexp**, or alternatively, scan the QR code below. Your input is greatly appreciated.

I wish you all the best,

Maksim Sokolov
Toronto, August 2023.

Chapter 1

Mathematical expressions

1.1   Explanation

Mathematical expressions are our main tools when we construct a mathematical model. These expressions may use letters as placeholders for numbers. Consider the following expression:

$$2x + 10 \tag{1.1}$$

This expression can be read: "Take some number x, multiply[1] it by 2, and then add 10". The letter x in this expression may represent 2, 3, 3.45, or any other number that is permitted within the context of the model being created. Such a letter is called a *variable*.

Expression 1.1 gives different outputs based on different values of the variable x. If x is equal to 2, the expression is equal to 14:

$$2 \cdot 2 + 10 = 14$$

If x is equal to 3, the expression is equal to 16:

$$2 \cdot 3 + 10 = 16$$

If x is equal to 3.45, the expression is equal to 16.9:

$$2 \cdot 3.45 + 10 = 16.9$$

---

[1]Notice that the sign of multiplication "$\cdot$" (or "$\times$") is missing in the given expression. This is a usual way to imply multiplication in algebra, as long as multiplication is evident from the context. Sometimes, the sign "$\cdot$" must be used to avoid confusion. Indeed, if we intended to multiply 2 by 3, and wrote 23 instead of $2 \cdot 3$, this would look like the number 23.

More generally, a mathematical expression may involve values that are allowed to change and values that are fixed. All these values are connected by mathematical operations, such as addition, subtraction, division, multiplication and exponentiation. The values that are allowed to change are *variables*: they have that name because they can take *various* values. The fixed values fall into two categories: *coefficients*, which are multiplied by variables, and "free" *constants* (which are not multiplied by a variable). For instance, consider again Expression 1.1. This expression contains two terms. The first term, $2x$, represents the product of the coefficient 2 and the variable x. The second term, which is the number 10, is the constant. The variable x can take various values, whereas the coefficient 2 and the constant 10 are fixed, meaning they cannot change.

As was mentioned in the very beginning, mathematical expressions are very useful when we build a model of a situation. For example, Expression 1.1 could describe a car wash service pricing structure that includes a minimum price of \$10 for the first 5 minutes of the wash plus \$2 for each additional minute. Since the duration of each customer's wash is not known to the service provider beforehand, the additional time beyond the initial 5 minutes is represented by the variable x. Once the exact number of additional minutes becomes known, it is substituted into the expression, allowing the service provider to calculate the final price for the car wash.

Consider several more examples of expressions that involve multiple variables (note that variables of mathematical expressions can be denoted by uppercase or lowercase letters from various alphabets, as well as by indexed letters):

$$180 - \alpha - \beta \tag{1.2}$$

$$0.5at^2 \tag{1.3}$$

$$\frac{x_1 + x_2 + x_3 + \cdots + x_n}{n} \tag{1.4}$$

$$P(1+r)^n \tag{1.5}$$

Expression 1.2 computes the third angle of a triangle, given the other two angles α and β. For example, if the two given angles are 60° and 30°, the third angle is 90°:

$$180 - 60 - 30 = 90$$

Expression 1.3 calculates the distance (in meters) traveled by an object that starts from rest and accelerates at a constant rate of a meters per second squared, for t seconds. For example, an object that starts from rest and accelerates at 10 m/s^2 would have travelled 45 meters after 3 seconds:

$$0.5 \cdot 10 \cdot 3^2 = 45$$

Expression 1.4 computes the arithmetic average of values x_1, x_2, x_3, $\cdots$, x_n. For example, the average weight of four weights (in grams) $x_1 = 560$, $x_2 = 610$, $x_3 = 430$ and $x_4 = 520$, is equal to 530 grams:

$$\frac{560 + 610 + 430 + 520}{4} = 530$$

Expression 1.5 computes the amount of money collected after n years if P dollars is invested at an annual interest rate r (compounded annually). For example, \$100 will grow to \$121, if invested at 10% per year:

$$100(1 + 0.1)^2 = 121$$

As we have mentioned, the variables of mathematical expressions are allowed to take various values. There are expressions, however, that become undefined for some values of the variables. For example, consider the following expression:

$$2 + \frac{10}{x - 2}$$

We can substitute any number for x in this expression, except $x = 2$. This is because substituting 2 would result in division by zero, which is not a valid mathematical operation. Or, take a look at another example:

$$\sqrt{x - 3} + x^2$$

In this expression, the allowed values of x are those that are equal to or greater than 3. This is because any value of x less than 3 would result in taking the square root of a negative number, an operation that is not defined[2] within the context of the *real*[3] elementary algebra we are studying in this book.

It is possible to construct *composite* mathematical expressions, whereby an expression or expressions can be plugged into another expression. Consider, for example, the following expression:

$$A + 2B \tag{1.6}$$

---

[2]However, in general, it is possible to take even roots of negative numbers. This operation leads to what are known as *complex* numbers. There exists a profound and important theory concerning such numbers.

[3]That is, involving *real* (not complex) numbers.

In this expression, let A and B be expressions:

$$A = a + b$$

$$B = a^2 - b^2$$

Then, if $a = 2$ and $b = 3$, we have:

$$A = 2 + 3 = 5$$

$$B = 2^2 - 3^2 = -5$$

And expression 1.6 is equal to -5:

$$A + 2B = 5 + 2(-5) = -5$$

In other words, expression 1.6 can be written as:

$$a + b + 2(a^2 - b^2)$$

Mathematical expressions of a certain type are referred to as *algebraic expressions*. Simply put, these are mathematical expressions that always contain a variable, and are constructed by means of operations of addition, subtraction, multiplication, division, and taking rational[4] powers. A variable in an algebraic expression cannot be a part of the power.

For example, Expressions 1.2 - 1.4 are all algebraic expressions. Expression 1.5 is not an algebraic expression if n is treated as a variable. But if we fix n (which would happen in case we decided for how many years we want to invest money), the expression will become an algebraic expression. For example, if we decide that all our investments will be for 2.5 years and no other duration is going to be considered, then our expression will become:

$$P(1 + r)^{2.5}$$

This is now an algebraic expression with variables P and r.

Algebraic expressions represent the basic, yet profoundly significant, category of mathematical expressions encompassing variables. They constitute a logical progression in abstraction from arithmetic expressions. Studying algebraic expressions is a foundational step towards embarking on a broader exploration of algebra and mathematics in general. The focus of this book will revolve around algebraic expressions.

---

[4]Recall that a *rational* number is a number that has the form $\frac{m}{n}$, where m is an integer and n is a positive integer.

We will conclude this section by highlighting a significant point that that holds importance for the remainder of the book. In future chapters, you will discover that it is possible to transform one expression into another, an equivalent expression. It is crucial to understand that such equivalencies may not hold true for all variable values. To illustrate, consider the following equality:

$$\frac{x^2}{x} = x \tag{1.7}$$

In this case, the expression on the left side is undefined for $x = 0$ (because otherwise there would be division by 0), while the expression on the right side is defined for all values of x. Therefore, Equality 1.7 is valid for all x with the exception of $x = 0$. For $x = 0$, Equality 1.7 does not hold.

1.2   Examples of worked exercises

Example 1.1 *Consider the following algebraic expression:*

$$4xy + A(m - 2)$$

What does it equal to, if $x = 2$, $y = 5$, $A = 5$ and $m = -1$?

Solution.

$$4 \cdot 2 \cdot 5 + 5(-1 - 2) = 40 + 5(-3) = 40 - 15 = \boxed{25}$$

∎ **End of the example**

Example 1.2 *If $\alpha = 2$ and $\beta = 3$, evaluate the following expression:*

$$\frac{\alpha^\beta - \beta^\alpha}{\alpha\beta}$$

Solution.

$$\frac{2^3 - 3^2}{2 \cdot 3} = \frac{8 - 9}{6} = \boxed{-\frac{1}{6}}$$

∎ **End of the example**

Example 1.3 *Consider the following three expressions:*

$$A_1 = 2^m(4 - n) - m, \qquad A_2 = n^2(3 - m), \qquad \Omega = A_1 A_2^{A_1}$$

Evaluate Ω, if $m = 2$ and $n = 3$.

Solution. First let's resolve A_1 and A_2:

$$A_1 = 2^2(4 - 3) - 2 = 2, \qquad A_2 = 3^2(3 - 2) = 9$$

Now, we will plug these results into Ω:

$$\Omega = 2 \cdot 9^2 = \boxed{162}$$

∎ **End of the example**

Example 1.4 *Evaluate the algebraic expression*

$$v_1 + 2v_2 + 3v_3 + \cdots + nv_n$$

for $n = 5$, $v_1 = 3$, $v_2 = 2$, $v_3 = 4$, $v_4 = 1$ and $v_5 = 10$.

Solution.

$$3 + 2 \cdot 2 + 3 \cdot 4 + 4 \cdot 1 + 5 \cdot 10 = \boxed{73}$$

∎ **End of the example**

1.3 Exercises

In Exercises 1.1 to 1.20, evaluate expressions using the given values and *without* using a calculator.

Exercise 1.1

(a) $2xy + 3(x + y)$ $\qquad x = 2, \quad y = 3$

(b) $-x - y(3x - y)$ $\qquad x = -3, \quad y = -2$

(c) $2(x - y) - 3(x - 2y)$ $\qquad x = -4, \quad y = 2$

(d) $-y(x - y - 2z)$ $\qquad x = -2, \quad y = -3, \quad z = -4$

Exercise 1.2

(a) $-x - y - x^2 - y^2$ $\qquad x = -1, \quad y = -2$

(b) $xy(x^2 - y^2 - z^2)$ $\qquad x = -2, \quad y = -3, \quad z = -4$

(c) $x^2 - y^2 - x^2y^2z^2$ $\qquad x = 2, \quad y = -2, \quad z = -3$

(d) $(x - y)^2 - z(x^2 + y^2)$ $\qquad x = -2, \quad y = 3, \quad z = -3$

Exercise 1.3

$$-(-x + y^2 - z^3)^2 \qquad x = -2, \quad y = 2, \quad z = -1$$

Exercise 1.4

$$-\left(-x + \frac{x}{y}\right) \qquad x = -4, \quad y = -\frac{x}{2}$$

Exercise 1.5

$$(x^2 - y)(y^2 - x) \qquad x = -2, \quad y = x - 1$$

Exercise 1.6

$$(z^m - z^{m-1})zm \qquad\qquad m = 2, \quad z = -1$$

Exercise 1.7

(a) $x_1 x_2^2 x_3^3 x_4^4 \qquad x_1 = 1, \quad x_2 = -1, \quad x_3 = 2, \quad x_4 = -2$

(b) $\sqrt{y_1 + y_2 + y_3 + y_4} \qquad y_1 = -1, \quad y_2 = -2, \quad y_3 = -3, \quad y_4 = 10$

(c) $\sqrt{z_1\sqrt{z_2\sqrt{z_3}}} \qquad z_1 = 25, \quad z_2 = 8, \quad z_3 = 4$

Exercise 1.8

$$-\frac{1}{t^3}\left(s^3 + \frac{t^3}{s}\right) \qquad\qquad s = \frac{1}{2}, \quad t = -\frac{1}{2}$$

Exercise 1.9

$$\left(\frac{5x+7}{11-x} \cdot \frac{2+\alpha^2}{2-\beta^3}\right)^x \qquad\qquad \alpha = \frac{1}{2}, \quad \beta = -\frac{1}{2}, \quad x = 2$$

Exercise 1.10

$$\sqrt{A(1-A)(1-A^2)(1-A^3) - 11\phi + \phi^\phi} \qquad A = -2, \quad \phi = -A$$

Exercise 1.11

$$A = -2C\left(\frac{2B}{B-2} - \frac{C}{C+1}\right)$$

$$B = \frac{3x^2 - 2y}{z}, \quad C = \frac{4y}{z} + 1, \quad x = 1, \quad y = -2, \quad z = 2$$

Exercise 1.12

$$A = \frac{3A_1}{A_1 - 2} + \frac{6A_1 A_2}{A_2 + 2}$$

$$A_1 = \frac{xy}{z}, \quad A_2 = \frac{x+y}{z}, \quad x = 2, \quad y = -1, \quad z = -2$$

Exercise 1.13

$$A = \sqrt{35\left(\frac{2A_1^2 + 5A_1}{A_1 - 2} - \frac{A_2}{A_2 - A_1}\right)}$$

$$A_1 = \frac{n_1 n_5^2 + n_2 n_6}{n_3 n_7}, \qquad A_2 = \frac{n_4 n_6}{n_3 n_7} - 1,$$

$$n_1 = -1, \quad n_2 = 2, \quad n_3 = 1, \quad n_4 = -2, \quad n_5 = -1, \quad n_6 = 2, \quad n_7 = -2$$

Exercise 1.14

$$A = \frac{2\Gamma^2(\Gamma - 1)}{\Gamma - \beta}\sqrt{\frac{\beta + \delta}{\beta^2}}$$

$$\Gamma = \frac{\gamma\theta^3 + \eta\phi}{\zeta\kappa}, \qquad \beta = \omega\left(\frac{\delta\eta}{\zeta\kappa}\right)^2,$$

$$\delta = 1, \quad \eta = -2, \quad \gamma = -1, \quad \kappa = 1, \quad \omega = 3, \quad \phi = -2\gamma, \quad \theta = -1, \quad \zeta = -2$$

Exercise 1.15

$$A = \frac{(\beta^2 + \delta^2)\sqrt{\Theta - \Phi - 20.5}}{\zeta^2(\alpha^6 - 1)(\Delta^2 - 1)} + \frac{3\Theta}{\sqrt{\Omega} - \sqrt{\Xi}}$$

$$\Theta = \frac{2\epsilon + \zeta^2}{\gamma\delta}, \qquad \Phi = \frac{5\beta^3 - \alpha\sqrt{-\gamma}}{\delta}, \qquad \Delta = \frac{\gamma}{2} - \frac{\sqrt{\Xi}}{\sqrt{\Omega}},$$

$$\Omega = 9, \quad \Xi = 4, \quad \alpha = 2, \quad \beta = 3, \quad \delta = -2, \quad \epsilon = 1, \quad \gamma = -1, \quad \zeta = -2$$

Exercise 1.16

$$\frac{\alpha_1}{2} + \frac{\alpha_2}{3} + \frac{\alpha_3}{4} + \cdots + \frac{\alpha_n}{n + 1}$$

$$n = 6, \qquad \alpha_1 = 4, \qquad \alpha_2 = 9, \qquad \alpha_3 = 8, \qquad \alpha_4 = 10, \qquad \alpha_5 = 6, \qquad \alpha_6 = 14$$

Exercise 1.17

$$m_1 m_2 m_3 \cdots m_n$$

$$n = 5, \qquad m_1 = 1, \qquad m_2 = 2, \qquad m_3 = 3, \qquad m_4 = 4, \qquad m_5 = 5$$

Exercise 1.18

$$\frac{2^{n-1} z_1 z_2 z_3 \cdots z_n}{(z_1 - 1)^2 (z_2 - 1)^2 (z_3 - 1)^2 \cdots (z_n - 1)^2}$$

$$n = 6, \qquad z_1 = -1, \qquad z_2 = 2, \qquad z_3 = 2, \qquad z_4 = -1, \qquad z_5 = 2, \qquad z_6 = -1$$

Exercise 1.19

$$(A_1 + A_2^2 + A_3^3 + \cdots + A_n^n)^n$$

$$n = 5, \qquad A_1 = -2, \qquad A_2 = -1, \qquad A_3 = -1, \qquad A_4 = -1, \qquad A_5 = -1$$

Exercise 1.20

$$\frac{\Omega_1^m}{\Delta_1^n} + \frac{\Omega_2^m}{\Delta_2^n} + \frac{\Omega_3^m}{\Delta_3^n} + \cdots + \frac{\Omega_m^m}{\Delta_m^n}$$

$$m = 5, \qquad n = 3, \qquad \Omega_1 = -2, \qquad \Omega_2 = -1, \qquad \Omega_3 = -2, \qquad \Omega_4 = -1, \qquad \Omega_5 = 2$$

$$\Delta_1 = -2, \qquad \Delta_2 = -1, \qquad \Delta_3 = -2, \qquad \Delta_4 = -1, \qquad \Delta_5 = -1$$

$$* * * * * * *$$

Exercise 1.21 *A toy company sells two types of stuffed animals: bears and rabbits. It buys the bears at x dollars each and the rabbits at y dollars each. The company also pays \$2 commission for each toy sale. If the company sells a bears and b rabbits, write an algebraic expression to represent the total cost of the stuffed animals sold. Calculate the total cost, if $x = 15$, $y = 17$, $a = 150$ and $b = 230$.*

Exercise 1.22 *Suppose we have a square yard with a side length of s meters, and a square pool inside the yard with a side length of t meters, where t is less than s. The pool is centered in the yard, and we want to cover the area between the pool and the yard's perimeters with tiles that cost x dollars per square meter.*

Write an algebraic expression that represents the total price of the tiles necessary to complete this task. Evaluate this expression for $s = 11$, $t = 9$, and $x = 30$.

Exercise 1.23 *Suppose you are planning a pizza party. You can choose to order one large pizza with a radius of r inches, or two smaller pizzas with radii of r_1 and r_2 inches, respectively.*

Write an algebraic expression for the difference in area between one large pizza and two smaller pizzas, using the formula for the area of a circle: $A = \pi r^2$. For this problem, assume that π is equal to 3.14. Evaluate this expression for $r = 8$, $r_1 = 6$, and $r_2 = 5$, and determine which option is more cost-effective for the given values.

Exercise 1.24 *You are constructing a specific shape. Begin with a rectangular box that has dimensions x meters by y meters by z meters, where z is the smallest side. Then, take a cube with a side length equal to z and divide it into m equal parts. Next, attach n of these cube parts to the top of the box.*

Write an algebraic expression that represents the volume of this shape. Use the expression to find the volume of the shape for the following values: $x = 1$, $y = 0.8$, $z = 0.6$, $m = 4$ and $n = 2$.

Exercise 1.25 *A car is traveling at a constant speed of v kilometers per hour and has a fuel efficiency of f liters per 100 kilometers. The price of fuel is δ dollars per liter.*

Write an algebraic expression that represents the price of fuel consumed if the car will travel t hours. Evaluate this expression for $v = 80$, $f = 9$, $\delta = 1.5$ and $t = 3$.

1.4   Answers

1.1: (a) 27 (b) -11 (c) 12 (d) 27

1.2: (a) -2 (b) -126 (c) -144 (d) 64

1.3: -49

1.4: -2

1.5: 77

1.6: -4

1.7: (a) 128 (b) 2 (c) 10

1.8: -1

1.9: 4

1.10: 12

1.11: 19

1.12: -5

1.13: 4

1.14: -1

1.15: 10

1.16: 12

1.17: 120

1.18: -4

1.19: -32

1.20: -22

1.21: $ax + by + 2(a + b)$, \$6920

1.22: $x(s^2 - t^2)$, \$1200

1.23: $\pi r^2 - (\pi r_1^2 + \pi r_2^2)$ (or $\pi(r^2 - r_1^2 - r_2^2)$), $9.42\,in^2$, larger pizza option

1.24: $xyz + \frac{nz^3}{m}$, $0.588\ m^3$

1.25: $\frac{vtf\delta}{100}$, \$32.4

Chapter 2

Exponents and radicals

2.1 Explanation

For any numbers x and y, and positive numbers a and b, the following equalities are valid:

$$a^x a^y = a^{x+y} \tag{2.1}$$

$$\frac{a^x}{a^y} = a^{x-y} \tag{2.2}$$

$$(a^x)^y = a^{xy} \tag{2.3}$$

$$a^0 = 1 \tag{2.4}$$

$$a^{-x} = \frac{1}{a^x} \tag{2.5}$$

$$(ab)^x = a^x b^x \tag{2.6}$$

$$\left(\frac{a}{b}\right)^x = \frac{a^x}{b^x} \tag{2.7}$$

We will not be presenting strict proofs for these equalities in this section. However, let's walk through them for the sake of gaining a deeper understanding.

Take, for example, $x = 5$ and $y = 3$ in Equalities 2.1, 2.2 and 2.3. Then, Equality 2.1 states that:

$$a^5 a^3 = (aaaaa)(aaa) = \underbrace{aaaaaaaa}_{8 \ factors} = a^8$$

Equality 2.2 states that:

$$\frac{a^5}{a^3} = \frac{aaaaa}{aaa} = \frac{a}{a} \cdot \frac{a}{a} \cdot \frac{a}{a} \cdot \frac{a}{1} \cdot \frac{a}{1} = a^2$$

And Equality 2.3 states that:

$$(a^5)^3 = (aaaaa)^3 = (aaaaa)(aaaaa)(aaaaa) = \underbrace{aaaaaaaaaaaaaaa}_{15 \ factors} = a^{15}$$

Equality 2.4 follows from Equality 2.2:

$$1 = \frac{a^x}{a^x} = a^{x-x} = a^0$$

Equality 2.5 follows from Equalities 2.4 and 2.2:

$$\frac{1}{a^x} = \frac{a^0}{a^x} = a^{0-x} = a^{-x}$$

Let's take $x = 3$ in Equalities 2.6 and 2.7. Then 2.6 states that:

$$(ab)^3 = (ab)(ab)(ab) = (aaa)(bbb) = a^3 b^3$$

And Equality 2.7 states that:

$$\left(\frac{a}{b}\right)^3 = \frac{a}{b} \cdot \frac{a}{b} \cdot \frac{a}{b} = \frac{aaa}{bbb} = \frac{a^3}{b^3}$$

Note that in the equalities we are working with, the exponents x and y are not restricted to being whole numbers. We will now discuss the meaning of fractional exponents.

We will begin by reviewing an exponent expression of the form $b^{\frac{1}{n}}$, where n is a positive integer. For this, we first define the *n-th root of a number b* as a number r which, when it is raised to the power of n, equals b:

$$r^n = b$$

For example, the 5-th root of 32 is 2, because $2^5 = 32$. And the 5-th root of -32 is -2, because $(-2)^5 = -32$.

For even values of n, the definition of the n-th root may result in a positive number having two distinct n-th roots, one positive and one negative. For instance, since $3^4 = 81$ and $(-3)^4 = 81$, there are two fourth roots of 81: 3 and -3.

We now define a *principal n-th root* as an n-th root if n is odd, and the *positive* n-th root if n is even.

Thus, while 81 has two fourth roots, 3 and -3, it has only one principal fourth root, 3.

The principal n-th root of a number b is denoted as $b^{\frac{1}{n}}$. Hence, when we encounter exponent expressions of the form $b^{\frac{1}{n}}$, we work with principal n-th roots.

There is an equivalent way to denote the principal n-th root of a number b: $\sqrt[n]{b}$. The symbol $\sqrt{}$ is called "the radical symbol" or the "radix". Thus:

$$b^{\frac{1}{n}} = \sqrt[n]{b} \tag{2.8}$$

Here are several examples of using the radical symbol:

$$\sqrt[5]{32} = 2$$

$$\sqrt[5]{-32} = -2$$

$$\sqrt[4]{81} = 3 \quad (\textit{note that } \sqrt[4]{81} \neq -3)$$

You should already be well familiar with the second root of b, commonly referred to as the "square root" of b. The positive square root of b is denoted by $\sqrt{b}$. For instance, $\sqrt{16} = 4$ since $4^2 = 16$.

Additionally, it is worth noting that the third root of b is frequently referred to as the "cube root" of b.

Having reviewed exponent expressions of the form $b^{\frac{1}{n}}$, we can now move on to examining exponent expressions of the form $b^{\frac{m}{n}}$, where both m and n are positive integers. The expression $b^{\frac{m}{n}}$ is understood in one of the two equivalent ways (see Equality 2.3):

$$b^{\frac{m}{n}} = \left(b^{\frac{1}{n}}\right)^m$$

$$b^{\frac{m}{n}} = (b^m)^{\frac{1}{n}}$$

For example:

$$32^{1.2} = 32^{\frac{6}{5}} = \left(32^{\frac{1}{5}}\right)^6 = 2^6 = 64$$

$$81^{0.75} = 81^{\frac{3}{4}} = \left(81^{\frac{1}{4}}\right)^3 = 3^3 = 27$$

Raising to a negative power is covered by Equality 2.5. For example:

$$32^{-1.2} = \frac{1}{64}$$

In summary, we have explored the meaning and properties of numbers raised to rational powers. In this book, we will exclusively deal with rational powers.

While we presented Equalities 2.1-2.7 for positive numbers a and b, many of those equalities are valid for 0 and negative numbers. For example:

$$0^x 0^y = 0^{x+y} = 0$$

$$\frac{(-2)^3}{(-2)^2} = (-2)^{3-2} = (-2)^1 = -2$$

However, some of the presented equalities cannot be used for 0 or negative a and b. For example:

$$0^{-2} = \frac{1}{0^2} = \frac{1}{0}$$

Dividing by 0 is not allowed. Or consider:

$$(-4)^{\frac{3}{2}} = \left((-4)^{\frac{1}{2}}\right)^3$$

Although it is possible to calculate the square root of a negative number – which would result in what's known as a "complex number" – such computations are beyond the scope of the *real* algebraic expressions which are the subject of this book. As a result, we will consider even roots of negative numbers to be undefined.

If we use any number a (non-negative or negative) where it is allowed, there is a special case that we must mention:

$$\sqrt{a^2} = |a| \tag{2.9}$$

Here, $|a|$ is the *absolute value* of a. Let's define it formally:

$$|a| = \begin{cases} a, & \text{if } a \geq 0 \\ -a, & \text{if } a < 0 \end{cases}$$

For example, $|-3| = 3$ and $|3| = 3$. Equality 2.9 simply means that

$$\sqrt{3^2} = 3 \quad and \quad \sqrt{(-3)^2} = 3$$

Note that several of our exponent equalities can be easily extended for more than two variables. For example:

$$a^{x_1} a^{x_2} a^{x_3} \cdots a^{x_n} = a^{x_1 + x_2 + x_3 + \cdots + x_n} \tag{2.10}$$

$$\left(\left(\left(a^{x_1}\right)^{x_2}\right)^{x_3} \cdots\right)^{x_n} = a^{x_1 x_2 x_3 \cdots x_n} \tag{2.11}$$

$$(a_1 a_2 a_3 \cdots a_n)^x = a_1^x a_2^x a_3^x \cdots a_n^x \tag{2.12}$$

The final remark concerns algebraic expressions that contain powers of variables.

Algebraic expressions all variables of which are raised to whole powers (and are not a part of an expression raised to a fractional power) are called *rational algebraic expressions*. Given rational fixed values and rational values for the variables, such expressions return rational numbers. For example, the following algebraic expressions are rational:

$$3x^3 - 2x^{-2} + 4$$

$$\frac{4x^2y^2 - 3x + 4}{5x^2y + 4}$$

$$x(x^2 + y^2)^{-3}$$

Algebraic expressions that are not rational are called *irrational algebraic expressions*. Given rational fixed values and rational values for the variables, such expressions may return irrational numbers[1]. For example, the following expressions are irrational:

$$2\sqrt{x} + x^3$$

$$\frac{3x^2 + 2x + 1}{3x^{\frac{3}{4}} - 1}$$

$$\sqrt{2x + 3} + \sqrt{2x - 3}$$

---

[1]A number that is not rational, is called an *irrational number*. This means that it is not possible to find whole numbers m and n to represent an irrational number in the form $\frac{m}{n}$. An example of an irrational number is $\sqrt{2}$. Irrational numbers have an infinite number of decimal digits that do not follow a repeating pattern. This means that, for practical computations, an irrational number must always be approximated by a rational number. For example, $\sqrt{2}$ is approximately equal to 1.4142135624, rounded at the tenth decimal place.

2.2   Examples of worked exercises

Example 2.1 *Simplify the algebraic expression:*

$$(2z)^4 - 16(z^2)^2$$

Solution. 1. At first, we simplify the first term using Equality 2.6:

$$(2z)^4 = 2^4 z^4 = 16 z^4$$

2. Then, we simplify the second term using Equality 2.3:

$$16(z^2)^2 = 16 z^{2 \cdot 2} = 16 z^4$$

3. Subtracting the second term from the first, we obtain:

$$16 z^4 - 16 z^4 = 0$$

Thus:

$$(2z)^4 - 16(z^2)^2 = \boxed{0}$$

■ **End of the example**

Example 2.2 *Simplify and evaluate the algebraic expression for $x = 2, y = -2$:*

$$\frac{(2xy^2)^4}{(4x^2 y)^2}$$

Solution. It is important not to evaluate an expression for given values before it has been simplified, as doing so would require additional effort, particularly when dealing with complex expressions. However, it is crucial to recognize that during the simplification process, there is a possibility of losing some information (the information about the values where the expression is not defined). In the context of this example, it is necessary to highlight from the outset of the simplification process that the expression is undefined when $x = 0$ and $y = 0$.

1. First, simplify the numerator using Equalities 2.12 and 2.3:

$$(2xy^2)^4 = 2^4 x^4 (y^2)^4 = 2^4 x^4 y^{2 \cdot 4} = 16 x^4 y^8$$

2. Next, simplify the denominator again using Equalities 2.12 and 2.3:

$$(4x^2 y)^2 = 4^2 (x^2)^2 y^2 = 4^2 x^{2 \cdot 2} y^2 = 16 x^4 y^2$$

3. Now, return to the original expression. Using what we have obtained for the numerator and the denominator in steps 1 and 2, and using Equalities 2.2 and 2.4, we have:

$$\frac{16x^4y^8}{16x^4y^2} = \frac{16}{16} \cdot \frac{x^4}{x^4} \cdot \frac{y^8}{y^2} = 16^{1-1}x^{4-4}y^{8-2} = 16^0 x^0 y^6 = 1 \cdot 1 \cdot y^6 = y^6$$

Thus:

$$\frac{(2xy^2)^4}{(4x^2y)^2} = \boxed{y^6}$$

This equality is valid for all values x and y, except $x = 0$ and $y = 0$.

4. We see that the final expression does not depend on x (but cannot be used to evaluate the original expression for $x = 0$), so we will not need to use one of the given values $x = 2$. Evaluating the expression using $y = -2$ yields:

$$y^6 = (-2)^6 = \boxed{64}$$

■ **End of the example**

Example 2.3 *Simplify and evaluate the algebraic expression for $m = 2, n = -1$:*

$$\frac{\left(\left((2m)^{-2}\right)^3\right)^{-2}}{m^{-8}(2m)^{14}} + \frac{(3n)^2}{3n^5}$$

Solution. 1. Simplifying the numerator of the first fraction using Equalities 2.11 and 2.6:

$$\left(\left((2m)^{-2}\right)^3\right)^{-2} = (2m)^{(-2)3(-2)} = (2m)^{12} = 2^{12}m^{12}$$

2. Simplifying the denominator of the first fraction using Equalities 2.6 and 2.1:

$$m^{-8}(2m)^{14} = m^{-8}2^{14}m^{14} = 2^{14}m^{-8}m^{14} = 2^{14}m^{-8+14} = 2^{14}m^6$$

3. Simplifying the numerator of the second fraction using Equality 2.6:

$$(3n)^2 = 3^2 n^2 = 9n^2$$

4. Returning to the main expression, plugging the simplified expressions we found in steps 1-3, and using Equalities 2.2 and 2.5:

$$\frac{2^{12}m^{12}}{2^{14}m^6} + \frac{9n^2}{3n^5} = 2^{12-14}m^{12-6} + 3n^{2-5} = 2^{-2}m^6 + 3n^{-3} = \frac{m^6}{2^2} + \frac{3}{n^3} = \frac{m^6}{4} + \frac{3}{n^3}$$

Thus:

$$\frac{\left(((2m)^{-2})^3\right)^{-2}}{m^{-8}(2m)^{14}} + \frac{(3n)^2}{3n^5} = \boxed{\frac{m^6}{4} + \frac{3}{n^3}}$$

Observe that the above equality is not valid for $m = 0$ and $n = 0$.

5. Evaluating the simplified expression for the given values:

$$\frac{m^6}{4} + \frac{3}{n^3} = \frac{2^6}{4} + \frac{3}{(-1)^3} = 16 - 3 = \boxed{13}$$

■ **End of the example**

Example 2.4 *Simplify and evaluate the algebraic expression for $\alpha = -2$ and $\beta = 4$:*

$$\left(\sqrt[3]{\alpha}\right)^6 + \left(\frac{\sqrt[3]{\beta^2}}{\sqrt[5]{\beta^3}}\right)^{30}$$

Solution. 1. We start with the first term. From Equality 2.8 we know that:

$$\sqrt[3]{\alpha} = \alpha^{\frac{1}{3}}$$

With this in mind, using Equality 2.3:

$$\left(\sqrt[3]{\alpha}\right)^6 = \left(\alpha^{\frac{1}{3}}\right)^6 = \alpha^{\frac{6}{3}} = \alpha^2$$

2. Working with the second term and using Equalities 2.3 and 2.2:

$$\left(\frac{\sqrt[3]{\beta^2}}{\sqrt[5]{\beta^3}}\right)^{30} = \left(\frac{(\beta^2)^{\frac{1}{3}}}{(\beta^3)^{\frac{1}{5}}}\right)^{30} = \left(\frac{\beta^{\frac{2}{3}}}{\beta^{\frac{3}{5}}}\right)^{30} = \left(\beta^{\frac{2}{3}-\frac{3}{5}}\right)^{30} = \left(\beta^{\frac{1}{15}}\right)^{30} = \beta^{\frac{30}{15}} = \beta^2$$

3. Adding the terms, we bring the original expression to the simplest form:

$$\left(\sqrt[3]{\alpha}\right)^6 + \left(\frac{\sqrt[3]{\beta^2}}{\sqrt[5]{\beta^3}}\right)^{30} = \boxed{\alpha^2 + \beta^2}$$

This equality is not valid for $\beta = 0$.

4. Evaluating for the given values:

$$(-2)^2 + 4^2 = \boxed{20}$$

■ **End of the example**

Example 2.5 *Simplify and evaluate first for $n = 2$ and then for $n = 3$:*

$$(-1)^n \left(\sqrt[n]{\sqrt[n]{2^n 3^n 4^n \cdots n^n}} \right)^{2n}$$

Solution. 1. Let's simplify the innermost radical, using 2.8 and 2.12:

$$\sqrt[n]{2^n 3^n 4^n \cdots n^n} = (2^n 3^n 4^n \cdots n^n)^{\frac{1}{n}} = 2^{\frac{n}{n}} 3^{\frac{n}{n}} 4^{\frac{n}{n}} \cdots n^{\frac{n}{n}} = 2 \cdot 3 \cdot 4 \cdot \cdots \cdot n$$

What we obtained is called "a factorial" of n, and is denoted as $n!$. That is:

$$1 \cdot 2 \cdot 3 \cdot 4 \cdots \cdot n = n!$$

2. Plugging the expression we obtained into the original expression, using 2.8 and 2.3, we get:

$$(-1)^n \left(\sqrt[n]{n!} \right)^{2n} = (-1)^n \left((n!)^{\frac{1}{n}} \right)^{2n} = (-1)^n (n!)^{\frac{2n}{n}} = \boxed{(-1)^n (n!)^2}$$

3. Evaluating for $n = 2$:

$$(-1)^2 (2!)^2 = (-1)^2 (1 \cdot 2)^2 = \boxed{4}$$

4. Evaluating for $n = 3$:

$$(-1)^3 (3!)^2 = (-1)^3 (1 \cdot 2 \cdot 3)^2 = -6^2 = \boxed{-36}$$

■ **End of the example**

2.3   Exercises

The variables in the exercises can take negative, zero, or positive numbers as long as the given expressions are well defined in the context of real-valued algebra (in each exercise assume such variables that their values cannot result in a zero denominator or an even root of a negative number).

Open the parentheses in Problems 2.1 - 2.5.

Exercise 2.1

(a) $(3x)^2$

(b) $(4y)^3$

(c) $(5z)^4$

(d) $(2\alpha)^3$

(e) $(3\beta)^4$

(f) $(4\gamma)^2$

(g) $(2\delta)^4$

(h) $(-5\epsilon)^4$

(i) $(-4A)^3$

(j) $(-2\eta)^3$

(k) $(-3\theta)^4$

(l) $(-4w)^2$

Exercise 2.2

(a) $(2xy)^5$

(b) $(-3yz)^4$

(c) $(-5xz)^4$

(d) $(5\alpha\beta)^4$

(e) $(-2vw)^3$

(f) $(3mn)^5$

(g) $(4\delta\epsilon)^2$

(h) $(-2st)^5$

(i) $(-3DF)^4$

(j) $(6BR)^3$

(k) $(5\alpha\phi\theta)^2$

(l) $(-4dsrw)^5$

Exercise 2.3

(a) $(x^2)^3$

(b) $(y^3)^4$

(c) $(z^4)^2$

(d) $(\alpha^2)^3$

(e) $(\beta^3)^4$

(f) $(\gamma^4)^2$

(g) $(\delta^2)^3$

(h) $(\epsilon^3)^4$

(i) $(W^4)^2$

(j) $((x^2)^3)^4$

(k) $((v^3)^4)^2$

(l) $((s^4)^2)^3$

Exercise 2.4

(a) $(2a^2b^3)^2$

(b) $(3c^2d^3e^4)^2$

(c) $(-2i^4j^3k^2)^2$

(d) $(4A^2B^3C^4)^2$

(e) $(-4o^3p^2q^4)^2$

(f) $(5\alpha^2\beta^3\gamma^4)^3$

(g) $(-5\delta^3\epsilon^2\zeta^4)^3$

(h) $(-3r^2s^3t^4u^2)^3$

(i) $(3l^2m^3n^4)^4$

(j) $(2a^2b^3)^4$

(k) $(-4f^3g^2h^4)^4$

(l) $(-3D^4E^3F^2G^4)^4$

Exercise 2.5

(a) $(x^{-2}y^3)^2$

(b) $(z^3w^{-1})^3$

(c) $(r^2s^{-3})^4$

(d) $(u^3v^{-2}w)^2$

(e) $(a^{-3}b^2)^3$

(f) $(c^{-1}d^3e^{-2})^2$

(g) $(-2x^2y^{-3})^2$

(h) $(3m^{-1}n^2)^3$

(i) $(-4p^3q^{-2})^3$

(j) $(2\alpha^2\beta^{-3})^4$

(k) $(-3S^3T^{-2}Q)^2$

(l) $(4\phi^{-1}\theta^2\omega^3)^{-1}$

Exercise 2.6 *Rewrite the following exponent expressions using radicals. An example of the expected answer:*

$$x^{\frac{5}{7}} + y^{1.75} = \sqrt[7]{x^5} + \sqrt[4]{y^7}$$

(a) $s^{\frac{2}{3}}$

(b) $t^{0.8}$

(c) $u^{0.75}$

(d) $w^{\frac{5}{3}}x^{\frac{2}{7}}$

(e) $y^{0.8}z^{0.5}$

(f) $\alpha^{\frac{3}{5}} + \beta^{\frac{7}{10}}$

(g) $\gamma^{\frac{6}{7}}\delta^{\frac{2}{3}}$

(h) $a^{0.5} + b^{0.75}$

(i) $\eta^{\frac{5}{6}}\theta^{\frac{1}{3}}$

(j) $l^{\frac{2}{5}}m^{\frac{3}{7}}$

(k) $\lambda^{\frac{4}{3}}\mu^{\frac{5}{8}}\nu^{\frac{2}{5}}$

(l) $G^{\frac{1}{4}} + H^{\frac{2}{3}} + J^{\frac{7}{10}}$

Exercise 2.7 *Rewrite the following expressions using fractional exponents. An example of the expected answer:*

$$\sqrt[5]{x^9}\sqrt[7]{y^5} = x^{\frac{9}{5}}y^{\frac{5}{7}}$$

(a) $\sqrt[5]{A}$

(b) $\sqrt[6]{y^5}$

(c) $\sqrt[7]{z^2}$

(d) $\sqrt[3]{q^2}$

(e) $\sqrt[4]{r^5}$

(f) $\sqrt[9]{u^2}$

(g) $\sqrt[3]{v}\sqrt[4]{w}$

(h) $\sqrt[5]{x^3} + \sqrt[3]{y^2}$

(i) $\sqrt[6]{w^7}\sqrt[2]{z}$

(j) $\sqrt[7]{\alpha^2}\sqrt[7]{\beta^3}\sqrt[7]{\gamma^4}$

(k) $\sqrt[3]{b^5} + \sqrt[5]{d^7} + \sqrt[7]{e^9}$

(l) $\sqrt[4]{c} + \sqrt[3]{d^2}\sqrt[2]{e^3}$

Exercise 2.8 *Open the parentheses and express the answer in the form of radicals:*

(a) $\left(a^{\frac{1}{2}}b^{\frac{1}{3}}\right)^2$

(b) $\left(\Delta^{\frac{1}{3}}\Gamma^{\frac{1}{2}}\right)^3$

(c) $\left(e^{\frac{1}{2}}f^{\frac{1}{3}}\right)^4$

(d) $\left(g^{\frac{1}{3}}h^{\frac{1}{2}}i^{\frac{1}{5}}\right)^2$

(e) $\left(-3j^{\frac{1}{3}}k^{\frac{1}{2}}\right)^3$

(f) $\left(2l^{\frac{1}{2}}m^{\frac{1}{3}}n^{\frac{1}{4}}\right)^2$

(g) $\left(-\alpha^{-\frac{1}{2}}\beta^{\frac{1}{2}}\right)^{\frac{2}{3}}$

(h) $\left(q^{\frac{1}{2}}r^{-\frac{1}{3}}\right)^{\frac{5}{3}}$

(i) $\left(-W^{\frac{1}{3}}Z^{-\frac{1}{2}}\right)^{\frac{7}{3}}$

(j) $\left(u^{-\frac{1}{2}}v^{\frac{1}{3}}\right)^{\frac{5}{2}}$

(k) $\left(-w^{-\frac{1}{3}}x^{\frac{1}{2}}y^{\frac{1}{4}}\right)^{-\frac{4}{3}}$

(l) $\left(\gamma^{-\frac{1}{2}}A^{-\frac{1}{3}}B^{\frac{1}{4}}\right)^{-\frac{5}{4}}$

Exercise 2.9 *Simplify the following expressions:*

(a) a^3a^2

(b) $c^{-5}c^3$

(c) $\alpha^3\alpha^{-2}$

(d) d^4d^{-6}

(e) E^2E^{-4}

(f) e^4e^{-1}

(g) $ll^{-1}g^{-2}g^5$

(h) $K^{-1}HH^{-3}K$

(i) $s^3k^3s^{-3}k^4$

(j) $j^{-1}j^{-4}j^{-2}$

(k) $\Delta^3\Delta^{-5}\Delta^3$

(l) $\lambda^4\lambda^2\lambda^{-5}$

Exercise 2.10 *Simplify the following expressions and state the answer in the form of radicals, where applicable:*

(a) $\Delta^{1.5}\Delta^{0.5}$

(b) $g^{-\frac{5}{3}}g^{\frac{2}{3}}$

(c) $R^{\frac{3}{4}}R^{-\frac{1}{4}}$

(d) $z^{\frac{4}{5}}z^{-\frac{6}{5}}$

(e) $p^{\frac{2}{3}}p^{-\frac{4}{3}}$

(f) $t^{\frac{4}{7}}t^{-\frac{1}{7}}$

(g) $u^{-\frac{2}{9}}u^{\frac{5}{9}}$

(h) $W^{\frac{1}{2}}W^{-\frac{3}{2}}$

(i) $Y^{\frac{3}{4}}Y^{\frac{1}{4}}$

(j) $Q^{-\frac{1}{5}}Q^{-\frac{4}{5}}Q^{-\frac{2}{5}}$

(k) $L^{1.5}L^{-2.5}L^{1.5}$

(l) $V^{\frac{4}{9}}V^{\frac{2}{9}}V^{-\frac{5}{9}}$

Exercise 2.11 *Simplify the following expressions and state the answer in the form of radicals:*

(a) $A^{\frac{5}{6}}A^{\frac{1}{3}}$

(b) $\beta^{-\frac{7}{4}}\beta^{\frac{2}{3}}$

(c) $W^{\frac{5}{8}}W^{-\frac{2}{5}}$

(d) $\delta^{\frac{3}{10}}\delta^{-\frac{7}{5}}$

(e) $m^{\frac{3}{4}}m^{-\frac{7}{2}}$

(f) $\zeta^{\frac{4}{7}}\zeta^{-\frac{1}{3}}$

(g) $G^{-\frac{2}{9}}G^{\frac{5}{6}}$

(h) $s^{\frac{1}{2}}s^{-\frac{3}{8}}$

(i) $I^{\frac{3}{4}}I^{\frac{3}{5}}$

(j) $\kappa^{-\frac{1}{5}}\kappa^{-\frac{4}{7}}\kappa^{-\frac{2}{9}}$

(k) $R^{\frac{3}{2}}R^{-\frac{5}{3}}R^{\frac{3}{4}}$

(l) $\lambda^{\frac{4}{9}}\lambda^{\frac{1}{7}}\lambda^{-\frac{5}{18}}$

Exercise 2.12 *Simplify the following expressions and state the answer in the form of radicals:*

(a) $\left(u^{\frac{1}{2}}\sqrt[3]{u}\right)^2$

(b) $\left(\sqrt{v}\sqrt[4]{v^3}\right)^3$

(c) $\left(w^{\frac{3}{2}}\sqrt[5]{w^2}\right)^{\frac{2}{3}}$

(d) $\left(\sqrt[3]{x^2}\sqrt{x}\right)^{\frac{4}{3}}$

(e) $\left(\alpha^{\frac{3}{2}}\sqrt[4]{\alpha^3}\right)^{\frac{5}{6}}$

(f) $\left(\beta^{\frac{1}{4}}\sqrt[5]{\beta^2}\right)^{\frac{8}{5}}$

(g) $\left(\sqrt[4]{\gamma^3}\gamma^{\frac{3}{2}}\right)^{\frac{7}{4}}$

(h) $\left(\sqrt{q}\sqrt[3]{q^5}\right)^{\frac{9}{10}}$

(i) $\left(\sqrt[5]{\omega^4}\omega^{\frac{2}{3}}\right)^{\frac{15}{8}}$

(j) $\left(\sqrt[3]{\kappa^{-5}}\sqrt[4]{\kappa^3}\sqrt[5]{\kappa^2}\right)^{-\frac{12}{5}}$

(k) $\left(\sqrt[5]{\mu^{-4}}\mu^{-\frac{1}{3}}\sqrt[5]{\mu^{-3}}\right)^{-\frac{15}{4}}$

(l) $\left(\left(\sqrt[3]{\nu^2}\sqrt[4]{\nu^5}\sqrt[3]{\nu^{-2}}\right)^{\frac{9}{4}}\right)^{\frac{4}{15}}$

Exercise 2.13 *Simplify the following expressions:*

(a) $\dfrac{(3aK^2)^3}{(6a^2K)^2}$

(b) $\dfrac{(4r^3G)^2}{(8r^2G^2)^2}$

(c) $\dfrac{(5T^4s^2)^3}{(10T^2s^4)^2}$

(d) $\left(\dfrac{xz^5y^3}{2z^3y^5x}\right)^3$

(e) $108\left(\dfrac{J^2K^2B^6}{K^26J^6B^2}\right)^3$

(f) $16\left(\dfrac{4H^3L^7}{H^7L^3}\right)^{-2}$

(g) $\dfrac{(3xy^{-2})^{-3}}{(3x^2y)^{-2}}$

(h) $36\dfrac{(-3r^{-3}G)^{-2}}{(2r^{-2}G^2)^2}$

(i) $\left(\dfrac{(T^4s^{-2})^3}{(2T^2s^4)^{-2}}\right)^3$

(j) $\dfrac{(A+B)^{-7}B^{-9}}{B^{-10}(A+B)^{-7}}$

(k) $\dfrac{(q+2p)^{-3}(q-3p)^4}{(q-3p)^3(q+2p)^{-3}}$

(l) $\left(\dfrac{(r-w)^3q^4}{(r-w)^2q^{-5}(r-w)}\right)^2$

Exercise 2.14 *Simplify the following expressions and state the answer in the form of radicals:*

(a) $\dfrac{\sqrt[3]{a^2}}{a^{\frac{5}{6}}}$

(b) $\dfrac{\sqrt[5]{L^3}L^{0.5}}{L^{0.75}}$

(c) $\dfrac{\sqrt[5]{M^4}M^{\frac{2}{3}}}{\sqrt[3]{M^7}}$

(d) $\dfrac{\sqrt[6]{\rho^5}\rho^{\frac{1}{3}}}{\rho^{0.3}\sqrt[3]{\rho^4}}$

(e) $\dfrac{\sqrt[3]{(K+2L)^2}\sqrt[4]{(K+2L)^3}}{(K+2L)^{\frac{3}{5}}(K+2L)^{\frac{5}{3}}}$

(f) $\dfrac{\sqrt[4]{(3\xi-\phi)^3}(3\xi-\phi)^{\frac{1}{2}}\sqrt[6]{(3\xi-\phi)^5}}{(3\xi-\phi)^{\frac{7}{9}}}$

(g) $\left(\dfrac{\sqrt[5]{Z^4}Z^{\frac{2}{3}}\sqrt[4]{Z^3}}{\sqrt[5]{Z^3}}\right)^{\frac{10}{97}}$

(h) $\sqrt[23]{\left(\dfrac{\pi^{\frac{35}{18}}}{\sqrt[6]{\pi^5}\pi^{\frac{1}{3}}\sqrt[3]{\pi^2}\sqrt[4]{\pi^3}}\right)^{12}}$

(i) $\left(\dfrac{\sqrt[3]{v^{0.4}}v^{-0.8}\sqrt[3]{v^{-2.5}}}{\sqrt[3]{v^{-0.75}}}\right)^{-2}$

(j) $\left(\dfrac{t^{-\frac{5}{6}}t^{\frac{2}{3}}}{\sqrt[4]{t^5}t^{-\frac{2}{3}}\sqrt[3]{t^2}\sqrt[4]{t^{-3}}t^{-\frac{5}{3}}}\right)^4$

Exercise 2.15 *Simplify the expression and evaluate it for $p = 60$, $q = 2$:*

$$\sqrt[6]{p^5}\;\sqrt[35]{\dfrac{\sqrt[6]{3}\left(((3p)^{-0.5})^3\right)^{-\frac{1}{3}}}{p^{-6}\sqrt[3]{(3p)^2}}}+\left(\dfrac{(2q)^{1.5}}{4q^4}\right)^{-2}$$

Exercise 2.16 *Simplify the expression and evaluate it for* $y = \frac{1}{128}$, $z = 2$:

$$\left(\sqrt[3]{4} \sqrt[3]{\frac{16}{\sqrt{\frac{y^{-2}}{z^3}}}} \sqrt{\frac{\sqrt[3]{y}\sqrt[4]{z^3}}{16y^{0.75}}} \right)^8$$

Exercise 2.17 *Simplify the expression and evaluate it for* $D = 4$, $P = 5$:

$$\left(\frac{\sqrt[6]{\frac{D^{-3}}{D^{1.25}\sqrt[4]{D^{-3}}}}}{D^4 \sqrt[4]{\frac{D^{3.5}}{D^{-\frac{3}{2}}\sqrt{D^3}\sqrt[3]{D^{-2}}}}} \right)^{-\frac{4}{9}} + \left(\sqrt[7]{\sqrt[3]{\sqrt[4]{P^5}}} \right)^{16.8}$$

Exercise 2.18 *Simplify the expression, given that* $n = 6$ *and* $m = 7$. *Then, evaluate it for* $\alpha = 64$ *and* $\beta = 2$:

$$\left(\sqrt{\sqrt[3]{\sqrt[4]{\cdots\sqrt[n]{\alpha}}}} \right)^{(n-1)!} + \sqrt[(m-1)!]{\left(\left((\beta^2)^3 \right)^4 \cdots \right)^m}$$

We remind that $k! = 1 \cdot 2 \cdot 3 \cdot 4 \cdots \cdot k$, *for any positive integer* k. *The operation* $k!$ *is called "k factorial".*

Exercise 2.19 *Simplify the expression, given that* $n = 6$ *and* $m = 3$. *Then, evaluate it for* $r = 9$ *and* $t = 3$:

$$\sqrt[2(n+1)]{r r^2 r^3 \cdots r^n} + t^{-1}t^2 t^{-3}t^4 \cdots t^{-(2m-1)}t^{2m}$$

Exercise 2.20 **(a)** *Simplify the expression:*

$$A = \left(\frac{\left(\sqrt[n]{x_1}\sqrt[n]{x_2}\sqrt[n]{x_3}\cdots\sqrt[n]{x_n} \right)^n}{\left(\left(\left((\Delta^{-x_1})^{-x_2} \right)^{-x_3} \cdots \right)^{-x_n} \right)} \right)^{\frac{1}{x_1 x_2 x_3 \cdots x_n}}$$

(b) *Given* $x_1 = 1, x_2 = 2, x_3 = 3, \cdots, x_n = n$, *find the expression for* B:

$$B = A^{n!}$$

(c) *Evaluate* B *for* $\Delta = 1$ *and* $n = 4$.
(d) *Evaluate* B *for* $\Delta = 2$ *and* $n = 3$.

2.4   Answers

2.1: (a) $9x^2$ (b) $64y^3$ (c) $625z^4$ (d) $8\alpha^3$ (e) $81\beta^4$ (f) $16\gamma^2$ (g) $16\delta^4$ (h) $625\epsilon^4$ (i) $-64A^3$ (j) $-8\eta^3$ (k) $81\theta^4$ (l) $16w^2$

2.2: (a) $32x^5y^5$ (b) $81y^4z^4$ (c) $625x^4z^4$ (d) $625\alpha^4\beta^4$ (e) $-8v^3w^3$ (f) $243m^5n^5$ (g) $16\delta^2\epsilon^2$ (h) $-32s^5t^5$ (i) $81D^4F^4$ (j) $216B^3R^3$ (k) $25\alpha^2\phi^2\theta^2$ (l) $-1024d^5s^5r^5w^5$

2.3: (a) x^6 (b) y^{12} (c) z^8 (d) α^6 (e) β^{12} (f) γ^8 (g) δ^6 (h) ϵ^{12} (i) W^8 (j) x^{24} (k) v^{24} (l) s^{24}

2.4: (a) $4a^4b^6$ (b) $9c^4d^6e^8$ (c) $4i^8j^6k^4$ (d) $16A^4B^6C^8$ (e) $16o^6p^4q^8$ (f) $125\alpha^6\beta^9\gamma^{12}$ (g) $-125\delta^9\epsilon^6\zeta^{12}$ (h) $-27r^6s^9t^{12}u^6$ (i) $81l^8m^{12}n^{16}$ (j) $16a^8b^{12}$ (k) $256f^{12}g^8h^{16}$ (l) $81D^{16}E^{12}F^8G^{16}$

2.5: (a) $\frac{y^6}{x^4}$ (b) $\frac{z^9}{w^3}$ (c) $\frac{r^8}{s^{12}}$ (d) $\frac{u^6w^2}{v^4}$ (e) $\frac{b^6}{a^9}$ (f) $\frac{d^6}{c^2e^4}$ (g) $\frac{4x^4}{y^6}$ (h) $\frac{27n^6}{m^3}$ (i) $-\frac{64p^9}{q^6}$ (j) $\frac{16\alpha^8}{\beta^{12}}$ (k) $\frac{9S^6Q^2}{T^4}$ (l) $\frac{\phi}{4\theta^2\omega^3}$

2.6: (a) $\sqrt[3]{s^2}$ (b) $\sqrt[5]{t^4}$ (c) $\sqrt[4]{u^3}$ (d) $\sqrt[3]{w^5}\sqrt[7]{x^2}$ (e) $\sqrt[5]{y^4}\sqrt{z}$ (f) $\sqrt[5]{\alpha^3}+\sqrt[10]{\beta^7}$ (g) $\sqrt[7]{\gamma^6}\sqrt[3]{\delta^2}$ (h) $\sqrt{a}+\sqrt[4]{b^3}$ (i) $\sqrt[6]{\eta^5}\sqrt[3]{\theta}$ (j) $\sqrt[5]{l^2}\sqrt[7]{m^3}$ (k) $\sqrt[3]{\lambda^4}\sqrt[8]{\mu^5}\sqrt[5]{\nu^2}$ (l) $\sqrt[4]{G}+\sqrt[3]{H^2}+\sqrt[10]{J^7}$

2.7: (a) $A^{\frac{1}{5}}$ (b) $y^{\frac{5}{6}}$ (c) $z^{\frac{2}{7}}$ (d) $q^{\frac{2}{3}}$ (e) $r^{\frac{5}{4}}$ (f) $u^{\frac{2}{9}}$ (g) $v^{\frac{1}{3}}w^{\frac{1}{4}}$ (h) $x^{\frac{3}{5}}+y^{\frac{2}{3}}$ (i) $w^{\frac{7}{6}}z^{\frac{1}{2}}$ (j) $\alpha^{\frac{2}{7}}\beta^{\frac{3}{7}}\gamma^{\frac{4}{7}}$ (k) $b^{\frac{5}{3}}+d^{\frac{7}{5}}+e^{\frac{9}{7}}$ (l) $c^{\frac{1}{4}}+d^{\frac{2}{3}}e^{\frac{3}{2}}$

2.8: (a) $a\sqrt[3]{b^2}$ (b) $\Delta\sqrt{\Gamma^3}$ (c) $e^2\sqrt[3]{f^4}$ (d) $\sqrt[3]{g^2}h\sqrt[5]{i^2}$ (e) $-27j\sqrt{k^3}$ (f) $4l\sqrt[3]{m^2}\sqrt{n}$ (g) $\sqrt[3]{\frac{\beta}{\alpha}}$ (h) $\frac{\sqrt[6]{q^5}}{\sqrt[9]{r^5}}$ (i) $-\frac{\sqrt[9]{W^7}}{\sqrt[6]{Z^7}}$ (j) $\frac{\sqrt[6]{v^5}}{\sqrt[4]{u^5}}$ (k) $\frac{\sqrt[9]{w^4}}{\sqrt[3]{x^2y}}$ (l) $\frac{\sqrt[8]{\gamma^5}\sqrt[12]{A^5}}{\sqrt[16]{B^5}}$

2.9: (a) a^5 (b) $\frac{1}{c^2}$ (c) α (d) $\frac{1}{d^2}$ (e) $\frac{1}{E^2}$ (f) e^3 (g) g^3 (h) $\frac{1}{H^2}$ (i) k^7 (j) $\frac{1}{j^7}$ (k) Δ (l) λ

2.10: (a) Δ^2 (b) $\frac{1}{g}$ (c) $\sqrt{R}$ (d) $\sqrt[5]{\frac{1}{z^2}}$ (e) $\sqrt[3]{\frac{1}{p^2}}$ (f) $\sqrt[7]{t^3}$ (g) $\sqrt[3]{u}$ (h) $\frac{1}{W}$ (i) Y (j) $\sqrt[5]{\frac{1}{Q^7}}$ (k) $\sqrt{L}$ (l) $\sqrt[9]{V}$

2.11: (a) $\sqrt[6]{A^7}$ (b) $\sqrt[12]{\frac{1}{\beta^{13}}}$ (c) $\sqrt[40]{W^9}$ (d) $\sqrt[10]{\frac{1}{\delta^{11}}}$ (e) $\sqrt[4]{\frac{1}{m^{11}}}$ (f) $\sqrt[21]{\zeta^5}$ (g) $\sqrt[18]{G^{11}}$ (h) $\sqrt[8]{s}$ (i) $\sqrt[20]{I^{27}}$ (j) $\sqrt[315]{\frac{1}{\kappa^{313}}}$ (k) $\sqrt[12]{R^7}$ (l) $\sqrt[42]{\lambda^{13}}$

2.12: (a) $\sqrt[3]{u^5}$ (b) $\sqrt[4]{v^{15}}$ (c) $\sqrt[15]{w^{19}}$ (d) $\sqrt[9]{x^{14}}$ (e) $\sqrt[8]{\alpha^{15}}$ (f) $\sqrt[25]{\beta^{26}}$ (g) $\sqrt[16]{\gamma^{63}}$ (h) $\sqrt[20]{q^{39}}$ (i) $\sqrt[4]{\omega^{11}}$ (j) $\sqrt[25]{\kappa^{31}}$ (k) $\sqrt{\mu^{13}}$ (l) $\sqrt[4]{\nu^3}$

2.13: (a) $\frac{3K^4}{4a}$ (b) $\frac{r^2}{4G^2}$ (c) $\frac{5T^8}{4s^2}$ (d) $\frac{z^6}{8y^6}$ (e) $\frac{B^{12}}{2J^{12}}$ (f) $\frac{H^8}{L^8}$ (g) $\frac{xy^8}{3}$ (h) $\frac{r^{10}}{G^6}$ (i) $64T^{48}s^6$ (j) B (k) $q-3p$ (l) q^{18}

2.14: (a) $\sqrt[6]{\frac{1}{a}}$ (b) $\sqrt[20]{L^7}$ (c) $\sqrt[15]{\frac{1}{M^{13}}}$ (d) $\sqrt[15]{\frac{1}{\rho^7}}$ (e) $\sqrt[20]{\frac{1}{(K+2L)^{17}}}$ (f) $\sqrt[36]{\frac{1}{(3\xi-\phi)^{47}}}$ (g) $\sqrt[6]{Z}$ (h) $\sqrt[3]{\frac{1}{\pi}}$ (i) $\sqrt{v^5}$ (j) $\frac{1}{t}$

2.15: $p+2q^5$, 124.

2.16: yz^7, 1.

2.17: $\sqrt{D^5} + P$, 37.

2.18: $\sqrt[6]{\alpha} + \beta^7$, 130.

2.19: $\sqrt{r^3} + t^3$, 54.

2.20: **(a)** $\dfrac{(x_1 x_2 x_3 \cdots x_n)^{\frac{1}{x_1 x_2 x_3 \cdots x_n}}}{\Delta^{(-1)^n}}$ **(b)** $\dfrac{n!}{\Delta^{(-1)^n} n!}$ **(c)** 24 **(d)** 384

Chapter 3

Distributive property

3.1   Explanation

It is clear that when we have an object and another identical object, we possess a total of two such objects. The same is true for variables: if we have a variable x and another same variable x, we have two of them:

$$x + x = 2x$$

This simply means that we can count variables. We can have x, $2x$, $3x$, etc. The fixed value that shows how many x's we have is called the *coefficient*. For example, in $3x$, 3 is the coefficient. We can have fractional coefficients (as in $1.25x$) and negative coefficients (as in $-3x$).

Suppose x represents the number of liters of water in a bucket. If Ken brings three and a half buckets of water ($3.5x$ liters) and Noor brings four buckets of water ($4x$ liters), they will collectively have 7.5 buckets of water or $7.5x$ liters:

$$3.5x + 4x = 7.5x$$

Of course, also:

$$7.5x - 4x = 3.5x$$

In general, for any (negative and non-negative) numbers a, b, and x, we have the following equality:

$$ax + bx = (a + b)x$$

Or:

$$(a + b)x = ax + bx \tag{3.1}$$

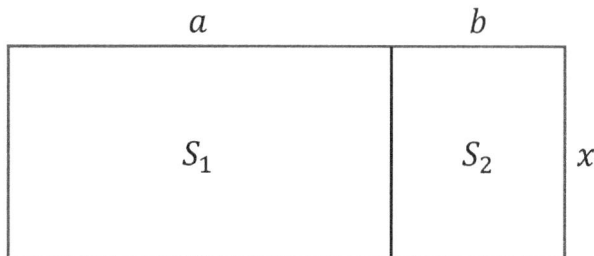

Figure 3.1: The geometric interpretation of the distributive property is depicted here. The area of the rectangle with side lengths $a + b$ and x is the sum of the areas of the smaller rectangles with side lengths a and x, and b and x.

Equality 3.1 is known as the "distributive property", because when the parentheses are opened, x is "distributed" to a and to b.

The distributive property can be understood geometrically. In Figure 3.1, the rectangle is divided into two smaller rectangles. Its area, which is equal to $(a + b)x$, is the sum of the areas $S_1 = ax$ and $S_2 = bx$. That is, we have Equality 3.1.

Equality 3.1 can be extended to a more general form. Given any number (or any expression) X and any numbers (or any expressions) A_1, A_2, A_3, ..., A_n, the following equality holds:

$$(A_1 + A_2 + A_3 + \cdots + A_n)X = A_1 X + A_2 X + A_3 X + \cdots + A_n X \qquad (3.2)$$

Here are examples of using Equality 3.2:

$$1.4x + 5x - 1.3x + 3.4x = (1.4 + 5 - 1.3 + 3.4)x = 8.5x$$

$$3(x + 2y) + (x^2 + y^2)(x + 2y) - (x + 2y) = (x + 2y)(3 + x^2 + y^2 - 1) = (x + 2y)(x^2 + y^2 + 2)$$

The distributive property holds significant value in algebra. In particular, this property leads to the process of "combining like terms". We consider terms in an algebraic expression as "like terms" if they differ in their coefficients or are identical. Here are several examples:

$$z^5 \text{ and } z^5 \text{ are like terms}$$

$$4z^5 \text{ and } 6z^5 \text{ are like terms}$$

$$-5v\sqrt[5]{s^7} \text{ and } 4.67v\sqrt[5]{s^7} \text{ are like terms}$$

$$4.5ght^3 \text{ and } -6.6ght^3 \text{ are like terms}$$

$$\frac{2r(x + 4)}{y} \text{ and } \frac{4r(x + 4)}{y} \text{ are like terms}$$

$$s^5 \text{ and } s^4 \text{ are NOT like terms}$$

$$wz^5 \text{ and } ws^5 \text{ are NOT like terms}$$

$$ght^3 \text{ and } g^3ht \text{ are NOT like terms}$$

"Like terms" can be combined by either adding them together or subtracting one from another as follows from the distributive property. Here are several examples:

$$2ght^3 + 15.6ght^3 = (2 + 15.6)ght^3 = 17.6ght^3$$

$$\frac{x+4}{y} - \frac{5(x+4)}{y} = (1-5)\frac{x+4}{y} = -\frac{4(x+4)}{y}$$

Finally, let's examine one more example:

$$A + 4B + 3A - 2B - 2A$$

In this expression, there are five terms. Among these five terms, we can identify two distinct groups. The first group contains three terms involving the variable A that differ only in their coefficients. The second group contains two terms involving the variable B that also differ only in their coefficients. In other words, we have separated the five terms into two groups of like terms. Now we can apply the distributive property (Equality 3.2) to each of these groups:

$$A + 4B + 3A - 2B - 2A = \underbrace{A + 3A - 2A}_{like\ terms} + \underbrace{4B - 2B}_{like\ terms}$$

$$= (1 + 3 - 2)A + (4 - 2)B = 2A + 2B$$

Of course, the variables A and B in the above expression may be algebraic expressions. For example, if $A = \frac{x^2}{2}$ and $B = \frac{y^3-1}{2}$, we can see that:

$$\frac{x^2}{2} + 4\left(\frac{y^3-1}{2}\right) + 3\left(\frac{x^2}{2}\right) - 2\left(\frac{y^3-1}{2}\right) - 2\left(\frac{x^2}{2}\right) = x^2 + y^3 - 1$$

3.2   Examples of worked exercises

Example 3.1 *Open the parentheses:*

$$-3x^2(x^3 - 2x)$$

Solution. Using Equality 3.1

$$-3x^2(x^3 - 2x) = (-3x^2)x^3 + (-3x^2)(-2x) = -3x^2x^3 + 6x^2x = \boxed{-3x^5 + 6x^3}$$

■ **End of the example**

Example 3.2 *Assuming that $q \geq 0$, simplify the expression:*

$$\sqrt{q}\left(q^{\frac{1}{2}} - q^{\frac{3}{2}} - 2q^2\right) - q\left(1 - \sqrt{q^3} - q + \sqrt{q}\right) + \sqrt{q^5}$$

Solution. 1. Open the first pair of parentheses, using Equality 3.2:

$$\sqrt{q}\left(q^{\frac{1}{2}} - q^{\frac{3}{2}} - 2q^2\right) = q^{\frac{1}{2}}\left(q^{\frac{1}{2}} - q^{\frac{3}{2}} - 2q^2\right) = q^{\frac{1}{2}}q^{\frac{1}{2}} - q^{\frac{1}{2}}q^{\frac{3}{2}} - q^{\frac{1}{2}}2q^2 = q - q^2 - 2q^{\frac{5}{2}}$$

2. Open the second pair of parentheses, again using Equality 3.2:

$$-q\left(1 - \sqrt{q^3} - q + \sqrt{q}\right) = -q\left(1 - q^{\frac{3}{2}} - q + q^{\frac{1}{2}}\right)$$

$$= -q + qq^{\frac{3}{2}} + qq - qq^{\frac{1}{2}} = -q + q^{\frac{5}{2}} + q^2 - q^{\frac{3}{2}}$$

3. The original algebraic expression becomes ("l.t." stands for "like terms"):

$$q - q^2 - 2q^{\frac{5}{2}} - q + q^{\frac{5}{2}} + q^2 - q^{\frac{3}{2}} + q^{\frac{5}{2}} = \underbrace{q - q}_{l.t.} \underbrace{-q^2 + q^2}_{l.t.} \underbrace{-2q^{\frac{5}{2}} + q^{\frac{5}{2}} + q^{\frac{5}{2}}}_{l.t.} - q^{\frac{3}{2}}$$

$$= (1-1)q + (-1+1)q^2 + (-2+1+1)q^{\frac{5}{2}} - q^{\frac{3}{2}} = -q^{\frac{3}{2}} = \boxed{-\sqrt{q^3}}$$

■ **End of the example**

Example 3.3 *Write as one fraction that contains no parentheses ($z \neq -1$, $z \neq 0$):*

$$\frac{z^2}{z+1} - \frac{3}{z^3}$$

To add or subtract fractions, both fractions must be transformed to ensure that they have the same denominator. To achieve this, we will multiply each fraction by 1, and will represent these 1's as suitable fractions:

$$\frac{z^2}{z+1}\cdot 1 - \frac{3}{z^3}\cdot 1 = \frac{z^2}{z+1}\cdot\frac{z^3}{z^3} - \frac{3}{z^3}\cdot\frac{z+1}{z+1} = \frac{z^2 z^3}{z^3(z+1)} - \frac{3(z+1)}{z^3(z+1)} = \frac{z^5 - 3(z+1)}{z^3(z+1)}$$

Now we will use the distributive property (Equality 3.1):

$$\frac{z^2 z^3 - 3(z+1)}{z^3(z+1)} = \boxed{\frac{z^5 - 3z - 3}{z^4 + z^3}}$$

■ **End of the example**

Example 3.4 *Factor* $-2x^2$ *out from the following expression:*

$$-8x^5 + 4x^3 - 2x^2$$

Solution. "Factoring $-2x^2$ out" means finding expressions a, b and c, such that:

$$-8x^5 + 4x^3 - 2x^2 = -2x^2(a + b + c)$$

That is, due to the distributive property:

$$-8x^5 + 4x^3 - 2x^2 = (-2x^2)a + (-2x^2)b + (-2x^2)c$$

We can find the unknown factors a, b and c by multiplying each term of the original expression by 1, written as fraction $\frac{(-2x^2)}{(-2x^2)}$ (in this operation we assume that $x \neq 0$):

$$-8x^5 + 4x^3 - 2x^2 = \frac{(-2x^2)}{(-2x^2)}\cdot(-8x^5) + \frac{(-2x^2)}{(-2x^2)}\cdot(4x^3) + \frac{(-2x^2)}{(-2x^2)}\cdot(-2x^2)$$

$$= (-2x^2)\cdot\frac{(-8x^5)}{(-2x^2)} + (-2x^2)\cdot\frac{4x^3}{(-2x^2)} + (-2x^2)\cdot\frac{(-2x^2)}{(-2x^2)}$$

$$= (-2x^2)4x^3 + (-2x^2)(-2x) + (-2x^2)1$$

Now we can factor $-2x^2$ out, using Equality 3.2:

$$(-2x^2)4x^3 + (-2x^2)(-2x) + (-2x^2)1 = \boxed{-2x^2(4x^3 - 2x + 1)}$$

Note that while we temporarily assumed that $x \neq 0$ to help us find the unknown factors, this condition is not really necessary for the obtained equality to hold at $x = 0$: both expressions of the equality are defined and the equality is valid for any value of x, including $x = 0$:

$$-8\cdot 0^5 + 4\cdot 0^3 - 2\cdot 0^2 = -2\cdot 0^2\cdot(4\cdot 0^3 - 2\cdot 0 + 1)$$

■ **End of the example**

Example 3.5 *Simplify the expression* $(a \neq 0, \ b \neq 0, \ ab \neq 0.5)$:

$$\frac{(2ab - 1)(3a^2b(a^2b^2 + 2) - 6a^2b)}{a^2b^2(2a^2b^2 - ab)}$$

Solution. 1. We notice that we can factor ab out in the denominator, using Equality 3.1:

$$2a^2b^2 - ab = ab(2ab - 1)$$

With this in mind, the original expression will become simpler:

$$\frac{(2ab - 1)(3a^2b(a^2b^2 + 2) - 6a^2b)}{a^2b^2(2a^2b^2 - ab)} = \frac{2ab - 1}{2ab - 1} \cdot \frac{3a^2b(a^2b^2 + 2) - 6a^2b}{a^2b^2ab}$$

$$= \frac{3a^2b(a^2b^2 + 2) - 6a^2b}{a^3b^3}$$

2. Next, we open the parentheses in the numerator using Equality 3.1:

$$3a^2b(a^2b^2 + 2) - 6a^2b = 3a^2b(a^2b^2) + 3a^2b(2) - 6a^2b = 3a^4b^3 + 6a^2b - 6a^2b = 3a^4b^3$$

3. Thus, the original expression becomes:

$$\frac{3a^4b^3}{a^3b^3} = \boxed{3a}$$

■ **End of the example**

Example 3.6 *Simplify the expression* $(k \neq 0)$:

$$\sqrt{\frac{k^3 + 2k^3 + 3k^3 + \cdots + nk^3}{nk + (n-1)k + (n-3)k + \cdots + k}}$$

Solution. 1. Let's rearrange the denominator by reversing the order of its terms:

$$nk + (n-1)k + (n-3)k + \cdots + k = k + 2k + 3k + \cdots + nk$$

2. Now we can use Equality 3.2 in the numerator and in the denominator:

$$\sqrt{\frac{k^3 + 2k^3 + 3k^3 + \cdots + nk^3}{k + 2k + 3k + \cdots + nk}} = \sqrt{\frac{k^3(1 + 2 + 3 + \cdots + n)}{k(1 + 2 + 3 + \cdots + n)}}$$

$$= \sqrt{\frac{k^3}{k} \cdot \frac{(1 + 2 + 3 + \cdots + n)}{(1 + 2 + 3 + \cdots + n)}} = \sqrt{k^2} = \boxed{|k|}$$

We remind the reader that $|k|$ is the absolute value of k (see Formula 2.9 and the explanation of that formula).

■ **End of the example**

3.3 Exercises

The variables in the exercises can take negative, zero, or positive numbers as long as the given expressions are well defined in the context of real-valued algebra (in each exercise assume such variables that their values cannot result in a zero denominator or an even root of a negative number).

Exercise 3.1 *Open the parentheses:*

(a) $3(w + 2)$

(b) $2(\alpha - 3)$

(c) $4(K + 1)$

(d) $(\mu - 2)5$

(e) $-(T + 4)$

(f) $(3z - 3)(-2)$

(g) $3(-3\rho + \beta - 4)$

(h) $-(-2Y - r - 1)$

(i) $(-4k - 2 - t)(-3)$

(j) $h(-5h^2 + h - 4)$

(k) $-7X\left(-2X - \frac{1}{X} - 1\right)$

(l) $\left(-\frac{7}{D} - \frac{2}{D^2} - D\right)(-3D)$

Exercise 3.2 *Open the parentheses:*

(a) $3b^2(4b^3 + 5b^4)$

(b) $4x^3(2x^2 + 3x^4)$

(c) $5D^4(3D^2 - 4D^3)$

(d) $(3\mu^2 + 4\mu^4)2\mu^3$

(e) $(2\nu^3 - 5\nu^2)3\nu^4$

(f) $-6E^2(2E^3 - 3E^4)$

(g) $-2z^4(-4z^2 - 5z^3)$

(h) $(-3G^3 + 2G^2)(-3G^4)$

(i) $-5\lambda^3(-4\lambda^2 - 3\lambda^4)$

(j) $-4H^3(-2K^4 - 5K^2)$

(k) $(3\alpha^3 + 4\alpha^2)(-2\phi^4)$

(l) $-3I^2(-5N^4 - 2M^3)$

Exercise 3.3 *Open the parentheses:*

(a) $\sqrt[4]{s^5}\left(\sqrt[6]{s^5} + \sqrt[8]{s^7}\right)$

(b) $\sqrt[5]{t^6}\left(\sqrt[4]{t^3} + \sqrt[7]{t^4}\right)$

(c) $\left(\sqrt[3]{u^4} + \sqrt[9]{u^5}\right)\sqrt[7]{u^3}$

(d) $-\sqrt[5]{v^7}\left(\sqrt[5]{v^3} - \sqrt[4]{v^5}\right)$

(e) $-\sqrt[7]{w^2}\left(-\sqrt[7]{w^5} - \sqrt[3]{w^4}\right)$

(f) $\left(\sqrt[5]{x^4} - \sqrt[8]{x^5}\right)\sqrt[9]{x^5}$

(g) $\left(-\sqrt[3]{y^5} - \sqrt[7]{y^3}\right)\sqrt[8]{y^5}$

(h) $\sqrt[5]{z^8}\left(z + \sqrt[10]{z^3} + \sqrt[7]{z^6}\right)$

Exercise 3.4   *Factor out the common factor without changing the order of the terms, making sure that the first term in the parentheses[1] has "+" sign. An example of the expected answer:*

$$-20x^5 - 5x^3 + 10x^2 = -5x^2(4x^3 + x - 2)$$

(a) $8p + 24$ **(b)** $5\beta - 15$ **(c)** $9A + 27$

(d) $6\delta - 18$ **(e)** $-3E - 9$ **(f)** $-f - 2$

(g) $20\gamma + 30\eta - 50$ **(h)** $-9H + 3\theta + 12$ **(i)** $12i + 16j - 8$

(j) $-4k^3 + 2k^2 - 6k$ **(k)** $-15L^4 - 5L^3 - 20L^2$ **(l)** $-10M^5 - 5N^4 - 30N^2$

Exercise 3.5   *Factor out the common factor without changing the order of the terms, making sure that the first term in the parentheses has "+" sign:*

(a) $14a^4b + 21a^5b$ **(b)** $10bc^6 + 15bc^8$ **(c)** $18p^5s - 12sp^6$

(d) $9qr^6 + 27qr^7$ **(e)** $16s^8p^3 - 8p^3s^7$ **(f)** $-20t^4k^5 + 30k^5t^5$

(g) $-28u^3v^2 + 14u^4v^2$ **(h)** $25w^4v^6 - 10v^5w$ **(i)** $-21w^7z^4 + 7w^9z^5$

(j) $12y^4x^3 - 18x^3y^5 - 6y^4x^5$ **(k)** $-8z^4\theta^3 - 16\theta^2z^4 - 8z^4\theta^5$ **(l)** $21\beta^4\alpha^2 - 7\alpha^2\beta^3 - 7\beta^9\alpha^2$

Exercise 3.6   *Factor $-t^4$ out of the following expressions, and use radicals in the answer where applicable. An example of the expected answer:*

$$t^{12} - t^{\frac{7}{5}} - 3t = -t^4\left(-t^8 + \sqrt[5]{\frac{1}{t^{13}}} + \frac{3}{t^3}\right)$$

(a) $1 + t$ **(b)** $t^3 - t^2$ **(c)** $-t^9 + t^3$

(d) $-2 - t - t^2$ **(e)** $t^8 - t^5 + t^4 - 1$ **(f)** $t^6 - 3t^4 + 2t^2$

(g) $2t^{\frac{8}{5}} - t^7 + t^{\frac{3}{4}} - 2$ **(h)** $-t^{12} - t^{\frac{9}{4}} + t^{\frac{2}{5}} - t$ **(i)** $-t^4 - t^{-\frac{5}{2}} + t^{-\frac{2}{3}}$

---

[1]The parentheses that will appear due to the factorization.

Exercise 3.7 *Factor $\sqrt[3]{x^7}$ out of the following expressions:*

(a) $\sqrt[3]{x^{10}} + \sqrt[3]{x^8}$

(b) $-\sqrt[3]{x^{11}} - \sqrt[4]{x^{15}}$

(c) $\sqrt[5]{x^{21}} - \sqrt{x^{11}}$

(d) $\sqrt[15]{x^{14}} + \sqrt[6]{x^7}$

(e) $-\sqrt[3]{x^{14}} - \sqrt[15]{x^{28}}$

(f) $-\sqrt[9]{x^{14}} - \sqrt[4]{x^{21}}$

(g) $\sqrt[4]{x^5} + \sqrt{x^5}$

(h) $\sqrt[5]{x^3} + \sqrt[4]{x^7}$

(i) $-\sqrt{x^5} - \sqrt[8]{x^3}$

Exercise 3.8 *Collect like terms for each of the following expressions:*

(a) $3a + 5a - 2a$

(b) $-7x + 3x^2 - 4x^2 + 2x$

(c) $4\beta + 2\gamma - 3\beta + \gamma$

(d) $5A^2 - 3A^3 + 2A^3 - 4A^2$

(e) $3\sqrt{p} - 4\sqrt{p^3} + 2\sqrt{p^3} + \sqrt{p}$

(f) $-2\sqrt[3]{y^5} + 5\sqrt[3]{y^5} - 3\sqrt[3]{y^5}$

(g) $3\alpha^2\beta - 2\alpha\beta^2 + \alpha^2\beta - 4\alpha\beta^2$

(h) $2\sqrt{x} + 3\sqrt[3]{x^2} - \sqrt{x} - \sqrt[3]{x^2}$

(i) $-\frac{1}{2}\sqrt[4]{m^3} - \frac{3}{4}\sqrt[4]{m^3} + \frac{1}{4}\sqrt[4]{m^3}$

(j) $4V^2 - \frac{1}{2}V^3 + \frac{3}{2}V^3 + 2V^2$

(k) $-\frac{1}{3}\sqrt[5]{n^4} + \frac{2}{3}\sqrt[5]{n^4} - \frac{1}{3}\sqrt[5]{n^4}$

(l) $3\omega^2 - 2\omega^3 + \omega^2 - 4\omega^3$

(m) $5\sqrt[3]{r^5} + 2\sqrt[5]{r^3} - 3\sqrt[3]{r^5} + \sqrt[3]{r^5}$

(n) $-3\theta^3 + 4\theta^2 - 2\theta^3 + 3\theta^2$

(o) $2\sqrt[4]{s^6} - \sqrt[4]{s^6} + \frac{1}{2}\sqrt[4]{s^6}$

(p) $-\frac{3}{4}\psi^4 + \frac{1}{4}\psi^4 - \frac{1}{2}\psi^4 + \frac{3}{4}\psi^4$

(q) $4\eta^3\phi^2 - 2\eta^3\phi^2 + \eta^3\phi^2 - 3\eta^3\phi^2$

(r) $-5\kappa\lambda + 3\kappa - 4\kappa\lambda - \kappa\lambda$

Exercise 3.9 *Apply the distributive property and collect like terms:*

(a) $-\left(3A^2B + 3\sqrt{C} + 2\sqrt[3]{D^4} - 5\sqrt[3]{D^4}\right) + A^2B - 2\sqrt{C} - 3A^2B + 4\sqrt{C} + 2\sqrt[3]{D^4} + \sqrt{C}$

(b) $-2\alpha\beta + \delta^3 - \alpha\beta - \delta^3 - \left(3\sqrt[3]{\gamma^2} + 5\sqrt[3]{\gamma^2} - 4\delta^3 + 2\delta^3 + 3\alpha\beta + 3\delta^3\right) - 2\sqrt[3]{\gamma^2}$

(c) $5\sqrt[3]{W^2} - \left(2X^2Y^3 + 3\sqrt[4]{Z} + 2\sqrt[3]{W^2} - 2\sqrt[4]{Z} - 3\sqrt[3]{W^2} + \sqrt[4]{Z} + X^2Y^3 - \sqrt[3]{W^2}\right)$

(d) $- \left(\sqrt[5]{r^4} - 2p^2q^3 + \sqrt[5]{r^4} - s^3 \right) - \left(5p^2q^3 - 2s^3 - 3p^2q^3 + \sqrt[5]{r^4} + 3s^3 - 3p^2q^3 \right)$

(e) $- \left(3G^2H - \sqrt{E} - 4\sqrt[4]{F^3} + G^2H + \sqrt[4]{F^3} \right) - \left(2\sqrt{E} - 3G^2H + 3\sqrt{E} + 5G^2H \right)$

(f) $- \left(x^3y - 2\sqrt[3]{u^4} \right) - \left(x^3y - 2v^2w \right) - \left(4\sqrt[3]{u^4} + \sqrt[3]{u^4} + 3v^2w - x^3y - v^2w + 5x^3y \right)$

(g) $- \left(3\Omega^3\Psi^2 - 2\sqrt[5]{N^3} + \Omega^3\Psi^2 \right) - \left(3\sqrt[5]{N^3} + \sqrt[5]{N^3} - 2\sqrt[3]{M^5} + 4\Omega^3\Psi^2 - \sqrt[3]{M^5} - 3\Omega^3\Psi^2 \right)$

(h) $3\delta^3 - \left(\alpha\beta + 2\delta^3 - 2\alpha\beta + \delta^3 \right) + 3\alpha\beta + 2\sqrt[3]{Q^2} - 3\sqrt[3]{Q^2} + 5\sqrt[3]{Q^2} - \left(\delta^3 - 4\delta^3 \right)$

(i) $- \left(\sqrt[3]{S^2} + X^2Y^3 \right) + 3\sqrt[3]{S^2} - \left(3\sqrt[4]{R} + X^2Y^3 \right) + 5\sqrt[3]{S^2} + 2\sqrt[4]{R} - \left(2X^2Y^3 - 2\sqrt[3]{S^2} \right)$

(j) $- W^3 - \left(3U^2V^3 + \sqrt[5]{T^4} - 2W^3 \right) - 2U^2V^3 - \left(\sqrt[5]{T^4} - 3U^2V^3 + \sqrt[5]{T^4} \right) + 3W^3$

Exercise 3.10 *Apply the distributive property and collect like terms:*

(a) $- 2x(x^3 - 3x + 1) - x^2(-3x^2 + 2) - x(x^3 + 4x - 2)$

(b) $3p(4p^2 - 5q + 1) - 2q(6p^2 + 7q) - p(12p^2 - 12pq - 15q + 3) - q(4 - 14q)$

(c) $- 5r(3r^2 + 2s - 4r^3) + 2s(4r^2 - 3r^3 + s) - r(20r^3 - 6r^2s - 15r^2 + 8rs - 10s)$

(d) $- 2V(6V^2 - 4V^3) + W(3V^2 - 2W^2) - V(8V^3 - 12V^2 - 3VW) - (6V^2W - 3W^3)$

(e) $- 6t(8t^2 + 9t^3) + u(3t^2 - 5u^2) - t^2(-54t^2 - 48t) - u(3t^2 - 5u^2 - t)$

(f) $c^2(3c + c(1 + 2c^3)) - c(3c^2 + (2c^3 + 1)c^2 - c)$

(g) $(((1+n)n)n)n - m(n^3(n + 2 + m)) - n^2(n^2 - m^3n^3 - 2mn) - m(n^5m^2 - mn^3 - n^4)$

(h) $\sqrt{b^3} \left(\sqrt{b^5} - \sqrt{b^3} + \sqrt[3]{b} \right) - \sqrt[6]{b^{11}} - b^3(b - 1)$

(i) $\sqrt{T^3} \left(\sqrt{T^3} \left(2\sqrt[3]{T^5} - 3\sqrt[4]{T^3} \right) \right) + T^3 \left(3\sqrt[4]{T^3} - 2\sqrt[3]{T^5} \right)$

(j) $\sqrt[3]{f^5}\left(2\sqrt[3]{f^4}-3\sqrt[4]{f^3}\right)-3\sqrt[4]{f^3}\left(\sqrt[4]{f^5}-\sqrt[3]{f^5}\right)-\left(2f^3-3\sqrt[4]{f^5}\left(\sqrt[4]{f^3}+1\right)\right)$

Exercise 3.11 *Write as one fraction without using parentheses. An example of the expected answer:*

$$\frac{x-y^2}{x^2+y}-\frac{x^2}{y}=\frac{-x^4-x^2y+xy-y^3}{x^2y+y^2}$$

(a) $\frac{x}{2}+\frac{x+y}{x}$

(b) $\frac{3}{y}-\frac{4}{x-y}$

(c) $\frac{c^2}{1+c}+\frac{5}{c^2}$

(d) $\frac{2a^2}{a+b}+\frac{3b}{a^2}$

(e) $-\frac{4p^2+p}{p-q}-\frac{3q}{p^2}$

(f) $\frac{3A^2+2B^3}{A+B}+\frac{A}{B}$

(g) $\frac{3x^2}{x+y}+\frac{4y}{x^2}-\frac{2x}{y^2}$

(h) $\frac{5m^2}{n}-\frac{7n}{m^2}-\frac{3mn}{m^2+n^2}$

Exercise 3.12 *Simplify the following expressions:*

(a) $\sqrt{\frac{x^3+x^2+x^2y^2}{x+y^2+1}}$

(b) $\frac{AB^2+4B^2}{A^2B+4AB}$

(c) $\frac{9x^2y-18xy^2}{9x^2-18xy}$

(d) $\frac{x^4y-8x^3y^2}{yx^4-4y^2x^3}-\frac{x-4y-4y}{x-y-3y}$

(e) $\frac{m^3n^2-m^2n^3}{m^4n^2-m^2n^2}\cdot\frac{m^2-1}{m-n}$

(f) $\frac{P^3Q^3-PQ^4}{P^2Q^3-PQ^4}-\frac{Q-PQ}{Q-P}$

(g) $\sqrt{\frac{2a^2b^2-4ab^3+2b^4}{2a^2-4ab+2b^2}}$

(h) $\sqrt{\frac{(t^2u)^2-t^3u^3-t^3u^3+(tu^2)^2}{t^2-2tu+u^2}}$

(i) $\sqrt[4]{\frac{v^2(u-2v)^4-u^2(u-2v)^4}{v^2-u^2}}$

(j) $\frac{wz+z+(wz+z)^3+(wz+z)^5}{z+z(wz+z)^2+z(wz+z)^4}$

Exercise 3.13 *Simplify the following expressions:*

(a) $\frac{a(a+b)}{a^2+ba}+\frac{6(x-z-t)}{3x-3t-3z}-\frac{mn-mn^2-mn^3}{mn(1-n^2-n)}$

(b) $\frac{y^5(6x^3-4y^2)}{x^2(xy^5+2y^6)}\cdot\frac{x^3+2x^2y}{2y^2-3x^3}$

(c) $\dfrac{f^2(3g(f+g^2f))}{f^3(g^3+g)} \cdot \dfrac{f(g^4-fg^3)-g^4}{g^3(-f^2+fg-g)}$

(d) $\dfrac{xw^2+xw}{x+w} \cdot \dfrac{x^2-x^2w}{xw(w-1)} \cdot \dfrac{w+x}{(w+1)x^2}$

(e) $1-(x+z)^5\left(\dfrac{(x+z)^5+y}{y(x+z)^5} - \dfrac{y}{(x+z)^5} - \dfrac{1}{y}\right)$

(f) $\left(\dfrac{xy}{z} + \dfrac{y}{z-1}\right)\dfrac{z(z-1)}{z-x+xz}$

(g) $\dfrac{rt-s(r+s)}{st^2}\left(\dfrac{r-s-t}{st} - \dfrac{r+s}{t^2} + \dfrac{1}{s} + \dfrac{1}{t}\right)^{-1}$

(h) $\dfrac{x(y+z)}{xy+xz} - \dfrac{z^{-2}}{x^2} + \dfrac{x^2+y^2}{(xyz)^2} - \dfrac{y^{-2}}{z^2}$

(i) $\dfrac{t^3(t+s-2)-s^2(t+3)}{s(s(-t-3)+t^3)+(t-2)t^3}$

(j) $\dfrac{G^5(T+2(H+3)^3)-GT^3(5(H+3)^3+G)}{G(G^3(2(H+3)^3+T)-T^3)-5(H+3)^3T^3}$

(k) $\dfrac{\alpha^2\beta(1-\alpha^2)-\alpha^3\beta^2(1-\beta^2)}{\alpha(\beta^2(\beta^2-1)-\alpha\beta)+\beta}\left(\dfrac{\alpha^2+\alpha(\beta+\beta^4)}{\beta^4+\alpha+\beta}\right)^{-1}$

(l) $\sqrt[3]{\left(\dfrac{n^2}{n+1} + \dfrac{n^3}{n^3+n^2}\right)^2} \cdot \sqrt{\dfrac{\sqrt[3]{n^4}\left(\sqrt{(20m+m^2)^3}+\sqrt[3]{n^5}\sqrt[3]{(20m+m^2)^2}\right)}{\sqrt[3]{(20m+m^2)^2n^2}\left(\sqrt[6]{(20m+m^2)^5}+\sqrt[3]{n^5}\right)}}$

Exercise 3.14 *Simplify* $\sqrt{A^{-1}B}$, *if:*

$$A = \dfrac{\sqrt{(uv)^3}\left(\sqrt{u}+\sqrt{v^3}-uv\left(\sqrt{u^3}+\sqrt{v}\right)\right)}{\left(\sqrt{u}+\sqrt{v^3}\right)\sqrt{(uv)^5}-\left(\sqrt{v}+\sqrt{u^3}\right)\sqrt{(uv)^7}} + u^{-2}+v^{-2}$$

$$B = \dfrac{u+v}{u} + \dfrac{v^2}{u^2}$$

Exercise 3.15 *Simplify the expression and evaluate it for $x = 8$, $y = 4$:*

$$\left(\frac{x^2}{y} - \frac{x^3 - y^2}{xy}\right)^{-1}\left(\frac{2x+y}{x+3y^2} - \frac{1}{xy} - \frac{y}{x}\right)\frac{(2y)^2 + x - y^2}{2\left(x - \frac{1}{2}y^{-1}\right) - 3yx^{-1}(y^2+1)}$$

Exercise 3.16 *Simplify the expression (in which b is an integer) and evaluate it for $a = 2$, $b = 3$:*

$$\left(\frac{ab\left(a\left(b\right)^2\right)^{-2}}{a+b+2} - b\left(\frac{a}{b}\right)^{-2} - \left(\frac{1}{a}\right)^{-1} + \frac{a\left(b^6 - 1\right) + (b+2)b^6}{a^2 b^{-1}(ab^4 + (b+2)b^4)}\right)^{2b}$$

Exercise 3.17 *Simplify $(AB)^3$ and evaluate it for $s = 2$, $t = 3$, if:*

$$A = \frac{\sqrt{t}}{\sqrt[3]{s}} - \frac{\sqrt{st} + \sqrt{t}}{\sqrt[3]{s^2}\sqrt[3]{t^2}} + \frac{\sqrt{s}\sqrt[3]{\frac{1}{t^2}} - \sqrt{st} + \sqrt[6]{\frac{1}{t}}}{\sqrt[3]{s^2}}$$

$$B = \frac{t^{\frac{4}{3}}}{s^{-1}\left(-\sqrt[6]{s} - \sqrt[6]{t^7}\right) + \left(t^{\frac{2}{3}} + 1\right)s^{-\frac{5}{6}}\sqrt{t}}$$

Exercise 3.18 *Simplify the expression:*

$$\sqrt{\frac{x + xy + x + 2xy + x + 3xy + \cdots + x + nxy}{nx^{-1} + y(x^{-1}n + x^{-1}(n-1) + x^{-1}(n-2) + \cdots + x^{-1})}}$$

Exercise 3.19 (a) *Simplify the expression:*

$$x(5 + x(4 + x(3 + x(2 + x(1 + x))))) - (x^5 + 2x^4 + 3x^3 + 4x^2 + 5x)$$

(b) *Simplify the generalized expression:*

$$x(n + x(n - 1 + \cdots + x(2 + x(1 + x))\cdots)) - (x^n + 2x^{n-1} + 3x^{n-2} + \cdots + nx)$$

Exercise 3.20 *Consider the following expressions:*

$$A = 1 + 2 + 3 + \cdots + n$$

$$A^* = n + (n-1) + (n-2) + \cdots + 1$$

Of course

$$A = A^*$$

(a) *Simplify $A + A^*$. Hint: Make the term-wise addition, then use the distributive property.*

(b) *Using the simplified expression you found in (a), show that:*

$$1 + 2 + 3 + \cdots + n = \frac{n(n+1)}{2}$$

(c) *Evaluate the following expression for $K = 2$ and $n = 1000$:*

$$\frac{K^2 + 2K^2 + 3K^2 + \cdots + nK^2}{Kn + K} + \frac{Kn}{2}$$

3.4 Answers

3.1: (a) $3w+6$ (b) $2\alpha-6$ (c) $4K+4$ (d) $5\mu-10$ (e) $-T-4$ (f) $-6z+6$ (g) $-9\rho+3\beta-12$ (h) $2Y+r+1$ (i) $12k+6+3t$ (j) $-5h^3+h^2-4h$ (k) $14X^2+7+7X$ (l) $21+\frac{6}{D}+3D^2$

3.2: (a) $12b^5+15b^6$ (b) $8x^5+12x^7$ (c) $15D^6-20D^7$ (d) $6\mu^5+8\mu^7$ (e) $6\nu^7-15\nu^6$ (f) $-12E^5+18E^6$ (g) $8z^6+10z^7$ (h) $9G^7-6G^6$ (i) $20\lambda^5+15\lambda^7$ (j) $8H^3K^4+20H^3K^2$ (k) $-6\phi^4\alpha^3-8\phi^4\alpha^2$ (l) $15I^2N^4+6I^2M^3$

3.3: (a) $\sqrt[12]{s^{25}}+\sqrt[8]{s^{17}}$ (b) $\sqrt[20]{t^{39}}+\sqrt[35]{t^{62}}$ (c) $\sqrt[21]{u^{37}}+\sqrt[63]{u^{62}}$ (d) $-v^2+\sqrt[20]{v^{53}}$ (e) $w+\sqrt[21]{w^{34}}$ (f) $\sqrt[45]{x^{61}}-\sqrt[72]{x^{85}}$ (g) $-\sqrt[24]{y^{55}}-\sqrt[56]{y^{59}}$ (h) $\sqrt[5]{z^{13}}+\sqrt[10]{z^{19}}+\sqrt[35]{z^{86}}$

3.4: (a) $8(p+3)$ (b) $5(\beta-3)$ (c) $9(A+3)$ (d) $6(\delta-3)$ (e) $-3(E+3)$ (f) $-(f+2)$ (g) $10(2\gamma+3\eta-5)$ (h) $-3(3H-\theta-4)$ (i) $4(3i+4j-2)$ (j) $-2k(2k^2-k+3)$ (k) $-5L^2(3L^2+L+4)$ (l) $-5(2M^5+N^4+6N^2)$

3.5: (a) $7a^4b(2+3a)$ (b) $5bc^6(2+3c^2)$ (c) $6p^5s(3-2p)$ (d) $9qr^6(1+3r)$ (e) $8s^7p^3(2s-1)$ (f) $-10t^4k^5(2-3t)$ (g) $-14u^3v^2(2-u)$ (h) $5v^5w(5vw^3-2)$ (i) $-7w^7z^4(3-w^2z)$ (j) $6x^3y^4(2-3y-x^2)$ (k) $-8z^4\theta^2(\theta+2+\theta^3)$ (l) $7\alpha^2\beta^3(3\beta-1-\beta^6)$

3.6: (a) $-t^4\left(-\frac{1}{t^4}-\frac{1}{t^3}\right)$ (b) $-t^4\left(-\frac{1}{t}+\frac{1}{t^2}\right)$ (c) $-t^4\left(t^5-\frac{1}{t}\right)$ (d) $-t^4\left(\frac{2}{t^4}+\frac{1}{t^3}+\frac{1}{t^2}\right)$ (e) $-t^4\left(-t^4+t-1+\frac{1}{t^4}\right)$ (f) $-t^4\left(-t^2+3-\frac{2}{t^2}\right)$ (g) $-t^4\left(-2\sqrt[5]{\frac{1}{t^{12}}}+t^3-\sqrt[4]{\frac{1}{t^{13}}}+\frac{2}{t^4}\right)$ (h) $-t^4\left(t^8+\sqrt[4]{\frac{1}{t^7}}-\sqrt[5]{\frac{1}{t^{18}}}+\frac{1}{t^3}\right)$ (i) $-t^4\left(1+\sqrt{\frac{1}{t^{13}}}-\sqrt[3]{\frac{1}{t^{14}}}\right)$

3.7: (a) $\sqrt[3]{x^7}\left(x+\sqrt[3]{x}\right)$ (b) $\sqrt[3]{x^7}\left(-\sqrt[3]{x^4}-\sqrt[12]{x^{17}}\right)$ (c) $\sqrt[3]{x^7}\left(\sqrt[15]{x^{28}}-\sqrt[6]{x^{19}}\right)$ (d) $\sqrt[3]{x^7}\left(\sqrt[5]{\frac{1}{x^7}}+\sqrt[6]{\frac{1}{x^7}}\right)$ (e) $\sqrt[3]{x^7}\left(-\sqrt[3]{x^7}-\sqrt[15]{\frac{1}{x^7}}\right)$ (f) $\sqrt[3]{x^7}\left(-\sqrt[9]{\frac{1}{x^7}}-\sqrt[12]{x^{35}}\right)$ (g) $\sqrt[3]{x^7}\left(\sqrt[12]{\frac{1}{x^{13}}}+\sqrt[6]{x}\right)$ (h) $\sqrt[3]{x^7}\left(\sqrt[15]{\frac{1}{x^{26}}}+\sqrt{\frac{1}{x^7}}\right)$ (i) $\sqrt[3]{x^7}\left(-\sqrt[6]{x}-\sqrt[24]{\frac{1}{x^{47}}}\right)$

3.8: (a) $6a$ (b) $-x^2-5x$ (c) $\beta+3\gamma$ (d) $-A^3+A^2$ (e) $4\sqrt{p}-2\sqrt{p^3}$ (f) 0 (g) $4\alpha^2\beta-6\alpha\beta^2$ (h) $\sqrt{x}+2\sqrt[3]{x^2}$ (i) $-\sqrt[4]{m^3}$ (j) V^3+6V^2 (k) 0 (l) $-6\omega^3+4\omega^2$ (m) $3\sqrt[3]{r^5}+2\sqrt[5]{r^3}$ (n) $-5\theta^3+7\theta^2$ (o) $\frac{3}{2}\sqrt[4]{s^6}$ (p) $-\frac{1}{4}\psi^4$ (q) 0 (r) $-10\kappa\lambda+3\kappa$

3.9: (a) $-5A^2B+5\sqrt[3]{D^4}$ (b) $-6\alpha\beta-\delta^3-10\sqrt[3]{\gamma^2}$ (c) $7\sqrt[3]{W^2}-3X^2Y^3-2\sqrt[4]{Z}$ (d) $-3\sqrt[5]{r^4}+3p^2q^3$ (e) $-6G^2H-4\sqrt{E}+3\sqrt[4]{F^3}$ (f) $-6x^3y-3\sqrt[3]{u^4}$ (g) $-5\Omega^3\Psi^2-2\sqrt[5]{N^3}+3\sqrt[3]{M^5}$ (h) $3\delta^3+4\alpha\beta+4\sqrt[3]{Q^2}$ (i) $9\sqrt[3]{S^2}-4X^2Y^3-\sqrt[4]{R}$ (j) $4W^3-2U^2V^3-3\sqrt[5]{T^4}$

3.10: (a) 0 (b) $-4q$ (c) $2s^2$ (d) W^3 (e) ut (f) c^2 (g) n^3 (h) 0 (i) 0 (j) $3\sqrt[4]{f^5}$

3.11: (a) $\frac{x^2+2x+2y}{2x}$ (b) $\frac{3x-7y}{yx-y^2}$ (c) $\frac{c^4+5c+5}{c^3+c^2}$ (d) $\frac{2a^4+3ab+3b^2}{a^3+a^2b}$ (e) $\frac{-4p^4-p^3+3q^2-3qp}{p^3-p^2q}$ (f) $\frac{3A^2B+2B^4+A^2+AB}{AB+B^2}$ (g) $\frac{3x^4y^2+4xy^3+4y^4-2x^4-2x^3y}{x^3y^2+x^2y^3}$ (h) $\frac{5m^6+5m^4n^2-7n^2m^2-7n^4-3m^3n^2}{m^4n+m^2n^3}$

3.12: (a) $|x|$ (b) $\frac{B}{A}$ (c) y (d) 0 (e) 1 (f) P (g) $|b|$ (h) $|tu|$ (i) $|u-2v|$ (j) $w+1$

3.13: (a) 2 (b) -2 (c) 3 (d) -1 (e) y (f) y (g) 1 (h) 1 (i) 1 (j) G (k) α (l) n

3.14: v

3.15: $\frac{x}{y}$, 2

3.16: a^{2b}, 64

3.17: $-s^2 t^2$, -36

3.18: $|x|$

3.19: **(a)** x^6 **(b)** x^{n+1}

3.20: **(a)** $n(1+n)$ **(b)** 2000

Chapter 4

Expansion and factoring

4.1 Explanation

Consider the expression $(a + b)(c + d)$, where a, b, c, d can be any numbers (non-negative or negative) or expressions.

To *expand* this expression (that is, to open the parentheses in this expression), we can use the distributive property (Equality 3.1) three times:

$$(a + b)(c + d) = a(c + d) + b(c + d) = ac + ad + bc + bd$$

This yields a well-known "FOIL" equality:

$$(a + b)(c + d) = ac + ad + bc + bd \tag{4.1}$$

The acronym FOIL stands for "First, Outer, Inner, Last" and it serves as a mnemonic for recalling Equality 4.1. Specifically, we must: find the product of the **F**irst terms inside each pair of parentheses (a and c), add the product of the **O**uter terms (a and d), add the product of the **I**nner terms (b and c), and finally add the product of the **L**ast terms (b and d).

Here is a simple example of using Equality 4.1:

$$(2 + x)(3 - x) = 2 \cdot 3 - 2x + 3x - x^2 = -x^2 + x + 6$$

Of course, Equality 4.1 can also be used for *factoring* (that is, for representing an expression as a product of two expressions). For example:

$$x^2 + xy^2 - xy - y^3 = xx - xy + xy^2 - y^2y = (x + y^2)(x - y)$$

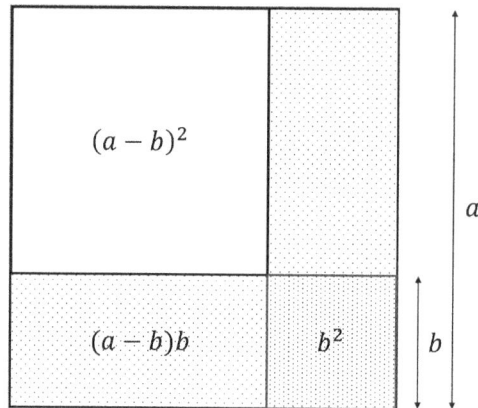

Figure 4.1: Geometric meaning of the equality $(a - b)^2 = a^2 - 2ab + b^2$.

Using Equality 4.1, it is not difficult to obtain other, very useful, equalities:

$$(a + b)^2 = a^2 + 2ab + b^2 \tag{4.2}$$

$$(a - b)^2 = a^2 - 2ab + b^2 \tag{4.3}$$

$$(a - b)(a + b) = a^2 - b^2 \tag{4.4}$$

$$(a + b)^3 = a^3 + 3a^2b + 3ab^2 + b^3 \tag{4.5}$$

$$(a - b)^3 = a^3 - 3a^2b + 3ab^2 - b^3 \tag{4.6}$$

$$(a + b)(a^2 - ab + b^2) = a^3 + b^3 \tag{4.7}$$

$$(a - b)(a^2 + ab + b^2) = a^3 - b^3 \tag{4.8}$$

The equalities that include $-b$ do not convey any fundamentally new information; they are provided merely for the sake of convenience. For instance, Equality 4.3 can be derived from Equality 4.2 by substituting $-b$ in place of b in Equality 4.2.

All equalities of this chapter are the main tools for expansion as well as for factoring expressions. Below are several examples:

$$(x^3 - 2)^2 = x^6 - 4x^3 + 4 \qquad \text{(Using Equality 4.2)}$$

$$27x^3 - 27x^2 + 9x - 1 = (3x - 1)^3 \qquad \text{(Using Equality 4.6)}$$

$$\left(xy - \sqrt{x}\right)\left(xy + \sqrt{x}\right) = x^2y^2 - x \qquad \text{(Using Equality 4.4)}$$

$$x^6 + 1 = (x^2 + 1)(x^4 - x^2 + 1) \qquad \text{(Using Equality 4.7)}$$

We will show how to obtain Equalities 4.2 and 4.5 and will leave the proof of the remaining equalities as an exercise. To prove Equality 4.2, we use Equality 4.1:

$$(a+b)^2 = (a+b)(a+b) = a^2 + ab + ba + b^2 = a^2 + 2ab + b^2$$

To prove Equality 4.5, we first use Equality 4.2 and then Equality 3.1:

$$(a+b)^3 = (a+b)^2(a+b) = (a^2 + 2ab + b^2)(a+b)$$
$$= (a^2 + 2ab + b^2)a + (a^2 + 2ab + b^2)b = a^3 + 2a^2b + b^2a + a^2b + 2ab^2 + b^3$$
$$= a^3 + 3a^2b + 3ab^2 + b^3$$

The equalities of this chapter have transparent geometric interpretation. For example, Figure 4.1 illustrates the geometric interpretation of Equality 4.3 in terms of areas:

$$(a-b)^2 = a^2 - (a-b)b - (a-b)b - b^2 = a^2 - ab - ab + b^2 = a^2 - 2ab + b^2$$

The image on the cover of the book illustrates Equality 4.5.

After verifying the validity of Equalities 4.2 - 4.8, it is important to memorize them, as they will be used frequently.

4.2 Examples of worked exercises

Example 4.1 *Simplify ($s \geq 0$):*

$$\frac{\left(\sqrt{s}+3\right)\left(2-\sqrt{s}\right)+\left(\sqrt{s}+\sqrt[4]{s}\right)^2}{\sqrt[4]{s^3}+3}$$

Solution. Using Equalities 4.1 and 4.2, we obtain:

$$\frac{\left(\sqrt{s}+3\right)\left(2-\sqrt{s}\right)+\left(\sqrt{s}+\sqrt[4]{s}\right)^2}{\sqrt[4]{s^3}+3} = \frac{2\sqrt{s}-s+6-3\sqrt{s}+s+2\sqrt[4]{s^3}+\sqrt{s}}{\sqrt[4]{s^3}+3}$$

$$= \frac{2\sqrt[4]{s^3}+6}{\sqrt[4]{s^3}+3} = \frac{2\left(\sqrt[4]{s^3}+3\right)}{\sqrt[4]{s^3}+3} = \boxed{2}$$

                      ■ **End of the example**

Example 4.2 *Simplify ($x \neq 0$):*

$$\frac{(x+2)^3-(x-2)^3}{(x+3)^2-(x-3)^2} - \frac{4}{3x}$$

Solution. Let's write this expression in the following way:

$$\frac{A}{B} - \frac{4}{3x}$$

where

$$A = (x+2)^3 - (x-2)^3$$
$$B = (x+3)^2 - (x-3)^2$$

1. Simplify A using Equalities 4.5 and 4.6:

$$(x+2)^3 - (x-2)^3 = x^3 + 6x^2 + 12x + 8 - (x^3 - 6x^2 + 12x - 8)$$
$$= x^3 + 6x^2 + 12x + 8 - x^3 + 6x^2 - 12x + 8 = 12x^2 + 16$$

2. Simplify B using Equalities 4.2 and 4.3:

$$(x+3)^2 - (x-3)^2 = x^2 + 6x + 9 - (x^2 - 6x + 9) = x^2 + 6x + 9 - x^2 + 6x - 9 = 12x$$

3. Return to the main expression:

$$\frac{A}{B} - \frac{4}{3x} = \frac{12x^2 + 16}{12x} - \frac{4}{3x} = \frac{12x^2 + 16}{12x} - \frac{16}{12x} = \frac{12x^2 + 16 - 16}{12x} = \frac{12x^2}{12x} = \boxed{x}$$

                      ■ **End of the example**

Example 4.3 *Assuming that $W \geq 0$, write the following expression in the form $(a+b)^2$:*

$$\sqrt{W}\left(\sqrt{W}+4\right) - (W+2)(W-2) + W^2$$

Solution. 1. Simplify the given expression using Equality 4.4:

$$\sqrt{W}\left(\sqrt{W}+4\right) - (W+2)(W-2) + W^2 = W + 4\sqrt{W} - (W^2 - 4) + W^2$$

$$= W + 4\sqrt{W} - W^2 + 4 + W^2 = W + 4\sqrt{W} + 4$$

2. Represent this expression in the form $(a+b)^2$ using Equality 4.2:

$$W + 4\sqrt{W} + 4 = \sqrt{W} + 2 \cdot 2\sqrt{W} + 2^2 = \boxed{\left(\sqrt{W}+2\right)^2}$$

■ **End of the example**

Example 4.4 *Factor the following expression in the form $(a+b)(c+d)$:*

$$k^2 - k - 6$$

Solution. We must have:

$$k^2 - k - 6 = (a+b)(c+d)$$

That is, according to Equality 4.1:

$$k^2 - k - 6 = ac + ad + bc + bd$$

Looking at the first term we see that we can take $a = c = k$. Using this information, for the remaining terms we have:

$$-k - 6 = kd + bk + bd$$

Looking carefully at the structure of the right side of the above equality, we rewrite the left side accordingly:

$$-3k + 2k - 3 \times 2 = kd + bk + bd$$

In other words, we see that $d = -3$ and $b = 2$. Therefore:

$$k^2 - k - 6 = \boxed{(k+2)(k-3)}$$

Another approach to solving this problem would be to use the distributive property three times:

$$k^2 - k - 6 = k^2 + 2k - 3k - 6 = k(k+2) - 3(k+2) = (k+2)(k-3)$$

■ **End of the example**

Example 4.5 *Simplify the expression* $(m \geq 0, \ m \neq 1, \ n \neq -2)$:

$$\frac{m(-n-2)+n+2}{n(n+4)+4} \cdot \frac{n+2}{(1-\sqrt{m})(1+\sqrt{m})}$$

Solution. Let's represent this expression in the following form:

$$\frac{A}{B} \cdot \frac{n+2}{C}$$

1. Work with A, first opening the parentheses and then using Equality 4.1:

$$m(-n-2)+n+2 = m(-n)-2m+n+2$$
$$= (-m)n+(-m)2+1n+1\cdot 2 = (-m+1)(n+2) = (1-m)(n+2)$$

2. Work with B, first opening the parentheses and then using Equality 4.2:

$$n(n+4)+4 = n^2+4n+4 = n^2+2\cdot 2n+2^2 = (n+2)^2$$

3. Work with C, opening the parentheses by using Equality 4.4:

$$(1-\sqrt{m})(1+\sqrt{m}) = 1-m$$

4. Return to the original expression:

$$\frac{A}{B} \cdot \frac{n+2}{C} = \frac{(1-m)(n+2)}{(n+2)^2} \cdot \frac{n+2}{1-m} = \frac{(1-m)(n+2)^2}{(1-m)(n+2)^2} = \boxed{1}$$

■ **End of the example**

4.3 Exercises

The variables in the exercises can take negative, zero, or positive numbers as long as the given expressions are well defined in the context of real-valued algebra (in each exercise assume such variables that their values cannot result in a zero denominator or an even root of a negative number).

Exercise 4.1 *Open the parentheses:*

(a) $(3 - x^2)(1 - x^2)$

(b) $(4 - y^2)(2 + y^2)$

(c) $(5\sqrt{z} - 2)(2\sqrt{z} + 3)$

(d) $(\alpha - 3\alpha^4)(\alpha + \alpha^4)$

(e) $\left(2 - \beta^{\frac{2}{3}}\right)\left(\beta^{\frac{4}{3}} + 4\right)$

(f) $\left(3\sqrt[3]{\gamma^5} - \sqrt[3]{\gamma}\right)\left(-2\sqrt[3]{\gamma^4} + 1\right)$

(g) $(2 - \rho^3)(3\rho^2 + \rho + 1)$

(h) $(1 - Y^2)(2Y + Y^3 - 1)$

(i) $(4k - k^2 - 2)(3k^2 + k - 1)$

(j) $(h - 5h^2 + h^3)(h^2 - h + 1)$

(k) $(-7X + 3)\left(2X^2 - X^{-1} - X^3\right)$

(l) $\left(-\frac{7}{D^2} - \frac{2}{D^3} + D\right)\left(3D^2 - D + 1\right)$

Exercise 4.2 *Open the parentheses using Equalities 4.2, 4.3, and then simplify:*

(a) $(5 + z)^2 - (1 + z)^2$

(b) $(6 + a)^2 - (2 - a)^2$

(c) $\left(7\sqrt{x} + 2\right)^2 + \left(3\sqrt{x} - 4\right)^2$

(d) $(\alpha + 2\alpha^2)^2 - (\alpha - 4\alpha^2)^2$

(e) $\left(\sqrt{V} + \sqrt{K}\right)^2 - \left(2\sqrt{V} - 2\sqrt{K}\right)^2$

(f) $(2\theta + \rho^2)^2 - (3\theta - 2\rho^2)^2$

(g) $(1 + s)^2 + (1 - s)^2 - (2 + s)^2$

(h) $(r^2 + r)^2 - (r^2 - r)^2 - (r^2 - 1)^2$

(i) $(1 + L)^4$

(j) $(2 + m)^4 - (2 - m)^4$

(k) $(1 - u)^4 - (u^2 - 2u)^2 - 2(u^2 - 2u)$

(l) $(1 - F)^6$

Exercise 4.3　*Open the parentheses using Equalities 4.2, 4.3 and 4.4, and then simplify:*

(a) $(x-2)(x+2)-(x+3)^2$

(b) $(\sqrt{q}-\sqrt{2p})^2-(\sqrt{q}-\sqrt{p})(\sqrt{q}+\sqrt{p})$

(c) $(2-f^2)(2+f^2)-(3-f^2)(f^2+3)$

(d) $(2z^2-\sqrt{z})\,(\sqrt{z}+2z^2)-(z^2+\sqrt{z})^2$

(e) $\left[\left(\sqrt{G}-\sqrt{2H}\right)\left(\sqrt{G}+\sqrt{2H}\right)\right]^2$

(f) $[(\beta^2+\beta^3)(\beta^2-\beta^3)]^2$

(g) $(1-h)(1+h)(1+h^2)$

(h) $(\sqrt{k}-\sqrt{2})(\sqrt{k}+\sqrt{2})(k+2)$

(i) $(2-z)(2+z)(4+z^2)-(4-z^2)^2$

(j) $(1-\theta)^4(1+\theta)^4-(\theta^2+1)^4$

(k) $(3-l-l^2)(3+l+l^2)+(l-l^2)^2$

(l) $\left(\sqrt{d^3}-\sqrt{d}+2\right)\left(\sqrt{d^3}-2+\sqrt{d}\right)$

Exercise 4.4　*Open the parentheses using Equalities 4.5, 4.6, 4.7, 4.8, and then simplify:*

(a)　$(d-2)^3-(d+3)^3$

(b)　$\left(\sqrt[3]{t}-\sqrt[3]{s}\right)^3-\left(\sqrt[3]{t}+\sqrt[3]{s}\right)^3$

(c)　$-(1+x)^3-(x-1)^3-(2-x)^3$

(d)　$(1-c)^3-(2-c)^3-(3+c)^3$

(e)　$(3+z)(9-3z+z^2)$

(f)　$(2-\sqrt{n})(4+2\sqrt{n}+n)$

(g)　$(\sqrt{k}+2f)(4f^2+k-2f\sqrt{k})-\sqrt{k^3}+f^3$

(h)　$(2-v)^2(v^2+2v+4)^2-64-v^6$

(i)　$(\gamma^3-\phi^3)((\gamma+\phi)^2-\gamma\phi)^{-1}$

(j) $(F^3 + 8)^3(F^2 + 4 - 2F)^{-3} - (2 - F)^3$

(k) $(\sqrt[4]{u} + \sqrt[4]{v})(u + \sqrt{uv} + v)(\sqrt[4]{u} - \sqrt[4]{v})$

(l) $(l + t)((l + t)^3 - 3lt(l + t))^{-1}(l^2 + t^2 - lt)$

Exercise 4.5 *Write the following expressions in the form $(a + b)(c + d)$. Use radicals in the answer, where applicable.*

An example of the expected answer:

$$-2x - 2(2x)^{0.5} + 3x^{0.5} + 3 \cdot 2^{0.5} = (3 - 2\sqrt{x})(\sqrt{2} + \sqrt{x})$$

(a) $h^2 + 3h + 2$ **(b)** $x^2 + 2x - 3$

(c) $t^2 - 5t + 6$ **(d)** $d - 7\sqrt{d} + 10$

(e) $fg + 2f - 2g^2 - 4g$ **(f)** $2rs + 2rt + s^2 + st$

(g) $-g + (3g)^{0.5} - g^{0.5} + 3^{0.5}$ **(h)** $-2x^5 - 6x^3 + 4x^2 + 12$

(i) $h^{57} - 2h^{34} - h^{23} + 2$ **(j)** $-R^{50} + R^{35} - R^{30} + R^{15}$

(k) $\sqrt[3]{D^7} - \sqrt[3]{D^5 Q} + \sqrt[3]{D^2}\sqrt[3]{Q} - \sqrt[3]{Q^2}$ **(l)** $3v^{1.75} + v^{2.5} + 2v^{0.75} + 6$

Exercise 4.6 *Write the following expressions in the form $(a+b)^2$ or $(a+b)^3$. Use radicals in the answer, where applicable.*

Examples of the expected answers:

$$-6(xy)^{0.5} + x + 9y = (\sqrt{x} - 3\sqrt{y})^2$$

$$64y^3 - 48\sqrt{3}y^2 + 36y - 3^{1.5} = (4y - \sqrt{3})^3$$

(a) $A^2 - 4A + 4$ **(b)** $9\mu^2 + 27\mu + \mu^3 + 27$

(c) $2\sqrt{zt} + z + t$ **(d)** $-3d^{2/3}s^{2/3} + 3d^{1/3}s^{4/3} + d - s^2$

(e) $27f^6 + 54f^4 + 36f^2 + 8$

(f) $S^{40} + 6S^{30} + 9S^{20}$

(g) $s^{1.25} + s^{2.5} + 0.25$

(h) $-24\phi^{2/3} + 8\phi + 24\phi^{1/3} - 8$

(i) $150\sqrt[3]{u^5} + 60\sqrt[3]{u^4} + 125u^2 + 8u$

(j) $16r^5 + 24r^3 + 9r$

(k) $6Q + 16Q^{0.75} + 10Q + 4\sqrt{Q}$

(l) $48f^{23/3} - 12f^{22/3} - 55f^8 + f^7 - (3f^4)^2$

(m) $(2 - \sqrt{x})(\sqrt{x} + 2) + 4(x + \sqrt{x}) - 2x$

(n) $(j^2 + k^2 + jk)(j^6 - 2j^3k^3 + k^6)(j - k)$

Exercise 4.7   *Simplify the following expressions and determine for which variables the equality of the original expression to the simplified expression is valid.*

   An example of the expected answer:

$$\frac{\left(\sqrt{x} + \sqrt{y}\right)^2}{\sqrt{xy}} - \sqrt{\frac{x}{y}} - \sqrt{\frac{y}{x}} = 2 \quad (for\ x > 0\ and\ y > 0)$$

(a) $\dfrac{1-s^2}{1-s}$

(b) $(\sqrt{p} + \sqrt{q})^2 - 2\sqrt{p}\sqrt{q}$

(c) $\dfrac{((W-2)^2 - (W+1)^2 + 7W)^2}{W+3}$

(d) $\dfrac{(\sqrt{\gamma+1}-1)^2 - \gamma - 2}{\sqrt{\gamma+1}}$

(e) $\dfrac{h^3 - g^3}{h - g} + gh$

(f) $\dfrac{K^3 + 12K^2 + 48K + 64}{(K+4)^2}$

(g) $\dfrac{(\sqrt{z} + \sqrt{z+1})^2 - 2z - 1}{\sqrt{z(z+1)}}$

(h) $\dfrac{2t - (\sqrt{t-1} - \sqrt{t+1})^2}{2\sqrt{t-1}\sqrt{t+1}}$

(i) $\dfrac{x^{\frac{9}{4}} + 3x^{\frac{7}{4}} + 3x^{\frac{5}{4}} + x^{\frac{3}{4}}}{x^{\frac{3}{2}} + 2x + x^{\frac{1}{2}}}$

(j) $\left(-\dfrac{(u+3)(u-3)}{u^2-9}\right)^p$

Exercise 4.8 *Simplify the following expressions:*

(a) $\dfrac{(3-n)^2}{(n-6)n+9}$

(b) $\left(\dfrac{\alpha(\alpha+1) + \alpha + 4}{\alpha^3 - 8}\right)^{-1}$

(c) $\dfrac{(3 - Q^3)(3 + Q^3) + 2(Q^3 - 1.5)^2 - 4.5}{(Q^3 - 3)^2}$

(d) $\dfrac{(\sqrt{x} + \sqrt{y})^3}{x + y + 2\sqrt{xy}}(\sqrt{x} - \sqrt{y})$

(e) $\dfrac{k(k - 2f) - 2f(\sqrt{2f} + \sqrt{k})(\sqrt{k} - \sqrt{2f})}{(k - 2f)^2}$

(f) $\dfrac{(g - h^4)^3(g + h^4)^2}{h^{16} + g^4 - 2g^2h^8}$

(g) $\dfrac{1}{a - b} + \dfrac{1}{a^2 + ab + b^2} - \dfrac{a + a^2 + b^2 - b + ab}{a^3 - b^3}$

(h) $\dfrac{(\Delta - 2)^4}{\Delta(\Delta(\Delta - 6) + 12) - 8}$

(i) $\dfrac{(p + 2)^{-1}}{(\sqrt{p} - \sqrt{2})(\sqrt{p} + \sqrt{2})}\left(\dfrac{(p - 2)^2(p + 2)^2}{p^2(48 + (p^2 - 12)p^2) - 64}\right)^{-1}$

(j) $\dfrac{(\sqrt{u + v} - \sqrt{u - v})^2 + 2\sqrt{u^2 - v^2}}{(0.25u + 1)^2 - (0.25u - 1)^2}$

(k) $\dfrac{s^2(s(s + 9) + 27) + 27s}{2s^2 + 12s + 18} - \dfrac{s^2((9 - s)s - 27) + 27s}{2s^2 - 12s + 18}$

(l) $\dfrac{3\sqrt[3]{x^5} - 3\sqrt[3]{x^4} - x^2 + x}{(\sqrt[3]{x} - \sqrt[3]{x^2})(\sqrt[3]{x^4} + \sqrt[3]{x^2} - 2x)}$

(m) $\left(\dfrac{A^2B + B^3}{AB} + \dfrac{A^2B + B^3}{A^2 + B^2}\right)\dfrac{A - B}{A^3 - B^3}$

(n) $\dfrac{c^2 + cd - 2d^2}{c^2 + 4cd + 4d^2} - \dfrac{3c}{2(c + 2d)} + \dfrac{1}{2}$

(o) $\dfrac{(x + 4)^5 - 5\sqrt{(x + 4)^5} + 6}{(x + 4)^5 - 6\sqrt{(x + 4)^5} + 9} - \dfrac{1}{\sqrt{(x + 4)^5} - 3}$

(p) $\dfrac{(3+s+t)^2 - (3-s-t)^2}{s+t}$

(q) $\dfrac{(2-r-2k)^2 - (2+r-2k)^2}{(1-r-k)^2 - (1+r-k)^2}$

(r) $\dfrac{(Z+Z^2+1)^3 - (Z-Z^2-1)^3}{(Z^2+1)(Z^4+5Z^2+1)}$

(s) $\dfrac{(1+f)(1-f)(1+2f) - (1-f)(1-2f)(1+2f)}{f(f(2f-1)-1)}$

(t) $\left(\dfrac{q^3}{q^3-8} + \dfrac{8(q-2)^{-2}}{q^2+2q+4}\right) \dfrac{q((q-2)q^2-8)+16}{q^3(q-2)+8}$

(u) $\dfrac{(w+w^4+w^5+w^8)^3}{w^{11}(w^4+3) + w^3(3w^4+1)}(w+1)^{-2}(w^2-w+1)^{-3}$

Exercise 4.9 *Simplify the expression:*

$$\frac{(x^2-2x+4)(\sqrt{x+3}+\sqrt{x+1})^2 - 2(x^3+8)}{(x-1)^2\sqrt{x^2+4x+3} + 3\sqrt{x^2+4x+3}}$$

Exercise 4.10 *Simplify the expression and evaluate it for $a = 2$:*

$$\left(\frac{(a^3-64)((a+4)^3 - 12a(a+4))}{a^3+64}\right)^3 \frac{(a^2-8\sqrt{a})^{-1}}{a^5+8a^3\sqrt{a}-64a^2-512\sqrt{a}}$$

Exercise 4.11 *Simplify the expression and evaluate it for $t = 0.5$:*

$$\frac{(1+t+t^2+t^3)^2(t+1)^{-2}(t^3+t) - ((t^2+3)t^2+3)t^3}{(1-t+t^2-t^3)^3(t-1)^{-2} - (t^3-t^2)(t^2(-t^2-3)-3)}$$

Exercise 4.12 *Simplify the expression and evaluate it for $t = -\frac{3}{4}$:*

$$\frac{\left(\sqrt{-x^2-2x}+\sqrt{x^2+2x+2}-\sqrt{2}\right)\left(\sqrt{1-(1+x)^2}+\sqrt{x^2+2x+2}+\sqrt{2}\right)}{\sqrt{-(x+1)^5+x+1}}$$

Exercise 4.13 *Simplify the expression and evaluate it for $u = 4$ and $v = 8$:*

$$\frac{(2u - v)^3 - u^2(6u - 13v)}{u^3 - uv^2} - \frac{(u - v)^2}{u^2 + uv} - \frac{(u + v)^2}{u^2 - uv}$$

Exercise 4.14 *Simplify the expression and evaluate it for $t = 0.5$ and $z = 0.25$:*

$$\frac{(z + t)^4 - (z^2 + t^2)^2}{(t^3 - z^3)((t - z)^2 - (t + z)^2)} - \frac{1}{z + t} + \frac{3t + z}{t^2 - z^2}$$

Exercise 4.15 *Simplify the expression:*

$$\frac{0.25}{x^2 + x + 1} \left(\sqrt{-\frac{x^3 - 1}{1 - x}} + \sqrt{-\frac{x^4 + x^3 - x - 1}{1 - x^2}} \right)^2$$

Exercise 4.16 *Simplify the expression:*

$$\left(\frac{1}{t - 1} + \frac{t + 1}{(t - 1)^2} + \frac{t^2 + 2t + 1}{(t - 1)^3} + \frac{t^3 + 3t^2 + 3t + 1}{(t - 1)^4} \right) \cdot \frac{(t - 1)^4}{t^2 + 1}$$

Exercise 4.17 *Simplify the expression:*

$$\left(A^B A^C \right)^D$$

where

$$A = 1 - \frac{\left(1 + \sqrt{2s}\right)(1 + 2s)}{\left(1 - \sqrt{2s}\right)^{-1}}, \quad B = \frac{\sqrt{s + \sqrt{s}}}{\sqrt{s - \sqrt{s}}}, \quad C = \frac{\sqrt{s - \sqrt{s}}}{\sqrt{s + \sqrt{s}}}, \quad D = \frac{\sqrt{s - 1}}{4\sqrt{s}}$$

Exercise 4.18 *Simplify the expression and evaluate it for $D = 3$ and $F = 4$:*

$$((\alpha^x \beta^x)^y)^{-2z}$$

where

$$\alpha = \sqrt[4]{\frac{F + D}{F^2 - D^2}}, \quad \beta = \sqrt[4]{\frac{1}{F - D}},$$

$$x = \frac{F^3 - D^3}{(F^{1.5} - D^{1.5})(\sqrt{FD} + D)}, \quad y = \frac{D^{2.5} - DF^{1.5}}{F^2 + DF + D^2}, \quad z = \frac{F + \sqrt{DF} + D}{D^2 - F\sqrt{DF}}$$

Exercise 4.19 *Simplify the expression:*

$$\sqrt[v]{\sqrt[w]{\xi^s}\,\sqrt[p]{\phi^q}}$$

where

$$\xi = \frac{(a + a^2 + 3)^2}{a(1 + a)} - \frac{a^4 + 2a^3 + a^2 + 27(a + a^2)^{-1}}{(a + 0.5)^2 + 2.75}$$

$$\phi = \frac{(b + 1)^4 - b(b - 2)^3 - (b - 2)^3}{b^3 + 1}$$

$$s = \frac{4ab(3a^2 - b^2)}{(\sqrt{12}a + 2b)(\sqrt{3}a - b)(3a^2 + b^2)}, \quad w = \frac{(a + b)^2 - (a - b)^2}{(a + b)^3 - (a - b)^3}$$

$$p = \left(\frac{1}{(a + 1)^3} - \frac{1}{(a^2 - 1)^3}\right) \cdot \frac{a^6 - 3a^4 + 3a^2 - 1}{a^3 - 3a^2 + 3a - 2}, \quad q = \frac{1}{a} - \frac{1 - a^3 b^3}{a(1 + ab)^2 - a^2 b}$$

$$v = \left(3 + 2\sqrt{b}\right)\left(a + 2\sqrt{b}\right) - 2(a + 3)\sqrt{b} - 3a$$

Exercise 4.20 *Simplify the expression and determine for which variables the equality of the original expression to the simplified expression is valid:*

$$(x - y)^{\frac{1}{256}}\,\sqrt[256]{\frac{(x + y)(x^2 + y^2)(x^4 + y^4)(x^8 + y^8)\cdots(x^{128} + y^{128})}{y^{256}}} + \frac{1}{x - y}$$

Exercise 4.21 **(a)** *Simplify $A - B$, if*

$$A = (1 + t)^2 + (2 + t)^2 + (3 + t)^2 + \cdots + (n + t)^2$$

$$B = (1 - t)^2 + (2 - t)^2 + (3 - t)^2 + \cdots + (n - t)^2$$

Hint: use the answer to Exercise 3.20(b).
(b) *Evaluate $A - B$ for $n = 10$ and $t = 2$.*

Exercise 4.22 **(a)** *Simplify:*
$$\frac{n^3}{3} - \frac{(n - 1)^3}{3} + n - \frac{1}{3}$$
(b) *Using part (a) of this exercise and Exercise 3.20(b), prove that:*

$$1^2 + 2^2 + 3^2 + \cdots + n^2 = \frac{n(n + 1)(2n + 1)}{6}$$

(c) *Evaluate $1^2 + 2^2 + 3^2 + \cdots + 12^2$.*

Exercise 4.23 **(a)** *Simplify:*

$$\frac{n^2(n+1)^2}{4} - \frac{(n-1)^2 n^2}{4}$$

(b) *Using part (a) of this exercise, prove that:*

$$1^3 + 2^3 + 3^3 + \cdots + n^3 = (1 + 2 + 3 + \cdots + n)^2 = \frac{n^2(n+1)^2}{4}$$

(c) *Evaluate $1^3 + 2^3 + 3^3 + \cdots + 10^3$.*

Exercise 4.24 **(a)** *Simplify $A - B$, if*

$$A = (1+d)^3 + (2+d)^3 + (3+d)^3 + \cdots + (n+d)^3$$

$$B = (1-d)^3 + (2-d)^3 + (3-d)^3 + \cdots + (n-d)^3$$

(b) *Evaluate $A - B$ for $n = 10$ and $d = 2$.*

Exercise 4.25 *Simplify $(A - B)C^{-1}$, if*

$$A = (1+r)^3 + (1+2r)^3 + (1+3r)^3 + \cdots + (1+nr)^3$$

$$B = (1-r)^3 + (1-2r)^3 + (1-3r)^3 + \cdots + (1-nr)^3$$

$$C = n^4 r^3 + 2n^3 r^3 + n^2 r^3 + 6n^2 r + 6nr$$

4.4 Answers

4.1: (a) $x^4 - 4x^2 + 3$ (b) $-y^4 + 2y^2 + 8$ (c) $10z + 11\sqrt{z} - 6$ (d) $-3\alpha^8 - 2\alpha^5 + \alpha^2$ (e) $2\beta^{\frac{4}{3}} - 4\beta^{\frac{2}{3}} - \beta^2 + 8$ (f) $-6\gamma^3 + 5\sqrt[3]{\gamma^5} - \sqrt[3]{\gamma}$ (g) $-3\rho^5 - \rho^4 - \rho^3 + 6\rho^2 + 2\rho + 2$ (h) $-Y^5 - Y^3 + Y^2 + 2Y - 1$ (i) $-3k^4 + 11k^3 - k^2 - 6k + 2$ (j) $h^5 - 6h^4 + 7h^3 - 6h^2 + h$ (k) $7X^4 - 17X^3 + 6X^2 - 3X^{-1} + 7$ (l) $3D^3 - D^2 + D + D^{-1} - 5D^{-2} - 2D^{-3} - 21$

4.2: (a) $8z + 24$ (b) $16a + 32$ (c) $58x + 4\sqrt{x} + 20$ (d) $12\alpha^3 - 12\alpha^4$ (e) $-3V + 10\sqrt{KV} - 3K$ (f) $-5\theta^2 + 16\theta\rho^2 - 3\rho^4$ (g) $s^2 - 4s - 2$ (h) $-r^4 + 4r^3 + 2r^2 - 1$ (i) $L^4 + 4L^3 + 6L^2 + 4L + 1$ (j) $16m^3 + 64m$ (k) 1 (l) $F^6 - 6F^5 + 15F^4 - 20F^3 + 15F^2 - 6F + 1$

4.3: (a) $-6x - 13$ (b) $3p - 2\sqrt{2pq}$ (c) -5 (d) $3z^4 - 2z^{5/2} - 2z$ (e) $G^2 - 4GH + 4H^2$ (f) $\beta^{12} - 2\beta^{10} + \beta^8$ (g) $-h^4 + 1$ (h) $k^2 - 4$ (i) $-2z^4 + 8z^2$ (j) $-8\theta^6 - 8\theta^2$ (k) $-4l^3 + 9$ (l) $d^3 - d + 4\sqrt{d} - 4$

4.4: (a) $-15d^2 - 15d - 35$ (b) $-6\sqrt[3]{st^2} - 2s$ (c) $-x^3 - 6x^2 + 6x - 8$ (d) $-c^3 - 12c^2 - 18c - 34$ (e) $z^3 + 27$ (f) $-\sqrt{n^3} + 8$ (g) $9f^3$ (h) $-16v^3$ (i) $\gamma - \phi$ (j) $2F^3 + 24F$ (k) $\sqrt{u^3} - \sqrt{v^3}$ (l) 1

4.5: (a) $(h+1)(h+2)$ (b) $(x-1)(x+3)$ (c) $(t-3)(t-2)$ (d) $(\sqrt{d}-2)(\sqrt{d}-5)$ (e) $(f-2g)(g+2)$ (f) $(2r+s)(s+t)$ (g) $(\sqrt{g}+1)(\sqrt{3}-\sqrt{g})$ (h) $(-2x^3+4)(x^2+3)$ (i) $(h^{34}-1)(h^{23}-2)$ (j) $(R^{20}+1)(R^{15}-R^{30})$ (k) $\left(\sqrt[3]{D^5} + \sqrt[3]{Q}\right)\left(\sqrt[3]{D^2} - \sqrt[3]{Q}\right)$ (l) $\left(2+\sqrt[4]{v^7}\right)\left(3+\sqrt[4]{v^3}\right)$

4.6: (a) $(A-2)^2$ (b) $(\mu+3)^3$ (c) $(\sqrt{z}+\sqrt{t})^2$ (d) $\left(\sqrt[3]{d} - \sqrt[3]{s^2}\right)^3$ (e) $(3f^2+2)^3$ (f) $(S^{20}+3S^{10})^2$ (g) $\left(\sqrt[4]{s^5}+0.5\right)^2$ (h) $(2\sqrt[3]{\phi}-2)^3$ (i) $\left(5\sqrt[3]{u^2}+2\sqrt[3]{u}\right)^3$ (j) $\left(4\sqrt{r^5}+3\sqrt{r}\right)^2$ (k) $\left(4\sqrt{Q}+2\sqrt[4]{Q}\right)^2$ (l) $\left(\sqrt[3]{f^7}-4\sqrt[3]{f^8}\right)^3$ (m) $(\sqrt{x}+2)^2$ (n) $(j^3-k^3)^3$

4.7: (a) $1+s$, $s \neq 1$ (b) $p+q$, $p, q \geq 0$ (c) $W+3$, $W \neq -3$ (d) -2, $\gamma > -1$ (e) $(g+h)^2$, $h \neq g$ (f) $K+4$, $K \neq -4$ (g) 2, $z > 0$ (h) 1, $t > 1$ (i) $\sqrt[4]{x^3} + \sqrt[4]{x}$, $x > 0$ (j) 1 or -1, $u \neq \pm 3$, p not an irreducible fraction with an even denominator.

4.8: (a) 1 (b) $\alpha - 2$ (c) 1 (d) $x - y$ (e) 1 (f) $g - h^4$ (g) 0 (h) $\Delta - 2$ (i) 1 (j) 2 (k) s^2 (l) 1 (m) $\frac{1}{A}$ (n) 0 (o) 1 (p) 12 (q) 2 (r) 2 (s) -3 (t) 1 (u) $w+1$

4.9: 2

4.10: $\frac{a^3-64}{a}$, -28

4.11: $\frac{t}{1-t}$, 1

4.12: $\frac{2}{\sqrt{1+x}}$, 4

4.13: $\frac{v}{u}$, 2

4.14: $\frac{1}{t-z}$, 4

4.15: 1

4.16: $4t$

4.17: $2s$

4.18: $F - D$, 1

4.19: 3

4.20: $\left|\dfrac{x}{y}\right|$, $x > y$, $y \neq 0$

4.21: (a) $2n(n + 1)t$ (b) 440

4.22: (a) n^2 (c) 650

4.23: (a) n^3 (c) 3025

4.24: (a) $2d^3n + dn(n + 1)(2n + 1)$ (b) 4780

4.25: 0.5

Chapter 5

Working with polynomials

5.1   Explanation

A distinctive class of algebraic expressions is known as *polynomials*. These expressions are constructed exclusively using addition, multiplication, and non-negative integer powers of variables. The variables within a polynomial are multiplied by coefficients, which may equal zero, be positive or negative numbers.

Here are several examples of polynomials:

$$5 \tag{5.1}$$

$$3x^2 \tag{5.2}$$

$$5x^2y^3 \tag{5.3}$$

$$-4z^2 + 2z \tag{5.4}$$

$$4.56x^3y^4 + 4x^2 \tag{5.5}$$

$$15.6x^5 + 4x^3 - 20 \tag{5.6}$$

$$-x^3y^7 + x^2 - y^2 \tag{5.7}$$

$$-6x^7 + 3x^5 - x^4 + 5x^2 - 10x - 5 \tag{5.8}$$

Expression 5.1 is a constant polynomial. Expressions 5.1, 5.2 and 5.3 are one-term polynomials, also known as *monomials*. Expressions 5.4 and 5.5 are two-term polynomials, also known as *binomials*. Expressions 5.6 and 5.7 are three-term polynomials, also known as *trinomials*. Expression 5.8 is a six-term one-variable polynomial. Expressions 5.3, 5.5 and 5.7 are two-variable polynomials.

The following are the examples of non-polynomials:

$$\frac{3}{x^2} + \frac{5}{y} \tag{5.9}$$

$$3x^{-4} + 3x^5 \tag{5.10}$$

$$\sqrt{xy} - 20x^2y \tag{5.11}$$

$$-3x^{\frac{3}{2}} - 5x^{\frac{5}{3}} + 7 \tag{5.12}$$

Expressions 5.9 and 5.10 are not polynomials because they contain negative powers of variables. Expressions 5.11 and 5.12 are not polynomials because they contain fractional powers of variables.

We will now discuss the concept of the *degree* of a polynomial. For a one-variable polynomial, identify the term with the highest power: that power is the degree of that polynomial. For a multi-variable polynomial, identify the term with the highest sum of powers: that sum is the degree of the polynomial.

Expression 5.1 is the polynomial of degree 0, because its only term contains a variable raised to power 0: $5x^0$. Expression 5.2 is the polynomial of degree 2. Expression 5.3 is the polynomial of degree 5. Expression 5.4 is the polynomial of degree 2. Expression 5.5 is the polynomial of degree 7. Expression 5.6 is the polynomial of degree 5. Expression 5.7 is the polynomial of degree 10. Expression 5.8 is the polynomial of degree 7.

In general, a one-variable polynomial of degree n has the following structure:

$$a_n x^n + a_{n+1} x^{n+1} + \cdots + a_1 x + a_0 \tag{5.13}$$

In this expression, a_1, a_2, ..., a_n represent coefficients, a_0 represents a constant, and x represents the variable[1]. It is noteworthy that in the general polynomial expression 5.13 the coefficients "behave" similarly to variables: they can be substituted with various numbers, enabling the construction of various polynomials. However, once a specific polynomial is created in accordance with 5.13, it will involve only the variable x. Thus, 5.13 is a model for the whole family of algebraic expressions that we call one-variable polynomials. Note that for such models (not limited to polynomials), there exists a convention of employing the first

---

[1]Recall that a *coefficient* is a fixed number that is multiplied by a variable and a *constant* is a fixed number that is not multiplied by a variable.

letters of the Latin alphabet (a, b, c,..., a_1, b_1, c_1,...) to represent the fixed values, while the final letters (x, y, z,..., x_1, y_1, z_1,...) are used to denote variables.

It is possible to perform various operations on polynomials.

Adding, subtracting, or multiplying polynomials is performed using the techniques we studied in the previous chapters: namely adding, subtracting, or multiplying algebraic expressions (see, for example, Exercises 4.1-4.4).

Adding one polynomial to another (or subtracting one polynomial from another) creates a new polynomial:

$$(3x^2 + 2x - 2) + (4x^4 - 4x^2 + x) = 4x^4 - x^2 + 3x - 2$$

$$(4x^2y^3 - 2x^2y^2 + 4xy) - (-8x^2y^3 - x^2y^2 - 2xy) = 12x^2y^3 - x^2y^2 + 6xy$$

Multiplying one polynomial by another creates a new polynomial:

$$(x^2 + 3)(2x^4 - x^2 + 1) = 2x^6 + 5x^4 - 2x^2 + 3 \tag{5.14}$$

Let's turn our attention to the division of polynomials. In certain instances, the division of polynomials can be straightforwardly performed, as exemplified by the following case:

$$\frac{x^2 - 9}{x + 3} = \frac{(x - 3)(x + 3)}{x + 3} = x - 3$$

In general, the task of dividing polynomials is not so straightforward. There is, however, a systematic method for performing the division, commonly referred to as the process of *long division*, which is similar to the process of long division of numbers. We will illustrate this process by carrying out the following division[2]:

$$\frac{2x^6 + 5x^4 - 2x^2 + 3}{x^2 + 3}$$

Step 1. Write the *dividend* (the numerator) and the *divisor* (the denominator) in the following way:

$$x^2 + 3 \overline{)\ 2x^6 + 5x^4 - 2x^2 + 3}$$

Ensure that all polynomials are written in the order of powers of their terms, starting from the term with the highest power and progressing to the terms with lower powers.

---

[2]Note Equality 5.14 which predicts the result to be $2x^4 - x^2 + 1$.

Step 2. Divide the first term of the dividend by the first term of the divisor, and place the result (which is $2x^4$) above the line, and above the term containing the same power:

$$
\begin{array}{r}
2x^4 \\
x^2 + 3 \overline{)\ 2x^6 + 5x^4 - 2x^2 + 3}
\end{array}
$$

Step 3. Multiply the divisor by the result that you have written above the line in Step 2. Write the resulting polynomial below the dividend, aligning the terms with the same powers. Change the sign of each term to its opposite, such that positive terms become negative and vice versa. Draw a line below the polynomial you have obtained.

$$
\begin{array}{r}
2x^4 \\
x^2 + 3 \overline{)\ 2x^6 + 5x^4 - 2x^2 + 3} \\
\underline{-\,2x^6 - 6x^4 }
\end{array}
$$

Step 4. Add the polynomial that has been written in Step 3 to the corresponding part of the dividend. Below the line, write the resulting polynomial, vertically aligning its terms with the terms of the same powers. Then "bring down" the next term from the dividend.

$$
\begin{array}{r}
2x^4 \\
x^2 + 3 \overline{)\ 2x^6 + 5x^4 - 2x^2 + 3} \\
\underline{-\,2x^6 - 6x^4 } \\
-\,x^4 - 2x^2
\end{array}
$$

Steps 5. The polynomial obtained in Step 4 now becomes the new dividend. Divide the first term of the new dividend by the first term of the divisor, and place the result (which is $-x^2$) above the line, and above the term (of the original dividend) containing the same power:

$$
\begin{array}{r}
2x^4 \ \ -\ x^2 \\
x^2 + 3 \overline{)\ 2x^6 + 5x^4 - 2x^2 + 3} \\
\underline{-\,2x^6 - 6x^4 } \\
-\,x^4 - 2x^2
\end{array}
$$

Steps 6. Multiply the divisor by the result you obtained in Step 5 (that is, by $-x^2$). Write the resulting polynomial below the new dividend, aligning the terms with the same powers. Change the sign of each term to its opposite. Draw a line below the obtained polynomial:

$$
\begin{array}{r}
2x^4 \quad\; - x^2 \\
x^2 + 3 \overline{)\, 2x^6 + 5x^4 - 2x^2 + 3} \\
-2x^6 - 6x^4 \\
\hline
-x^4 - 2x^2 \\
x^4 + 3x^2 \\
\hline
\end{array}
$$

Step 7. Add the polynomials above the line and then bring down the next term:

$$
\begin{array}{r}
2x^4 \quad\; - x^2 \\
x^2 + 3 \overline{)\, 2x^6 + 5x^4 - 2x^2 + 3} \\
-2x^6 - 6x^4 \\
\hline
-x^4 - 2x^2 \\
x^4 + 3x^2 \\
\hline
x^2 + 3
\end{array}
$$

Steps 8+. The polynomial obtained in Step 7 becomes the new dividend. Proceed with the process until there is nothing to bring down from the original dividend:

$$
\begin{array}{r}
2x^4 \quad\; - x^2 + 1 \\
x^2 + 3 \overline{)\, 2x^6 + 5x^4 - 2x^2 + 3} \\
-2x^6 - 6x^4 \\
\hline
-x^4 - 2x^2 \\
x^4 + 3x^2 \\
\hline
x^2 + 3 \\
-x^2 - 3 \\
\hline
0
\end{array}
$$

The result of a division of mathematical expressions is called the *quotient*. Specifically, the quotient of division of two polynomials is also known as a *rational algebraic expression*[3]. Hence, at the end of the demonstrated long division process, we have obtained the quotient, written above the original dividend:

$$
\frac{2x^6 + 5x^4 - 2x^2 + 3}{x^2 + 3} = x^4 - 4x^2 + 2
$$

---

[3]Recall that we already defined rational and irrational algebraic expressions in Chapter 2. But here we see an equivalent definition: rational algebraic expressions can be represented as a division of two polynomials, whereas irrational algebraic expressions cannot be represented in such a way.

Many students learn to perform the process of long division but often lack the understanding of what is occurring "behind the scenes". It is important to understand why the process works. To clarify this, we will examine the following division (having been assured first that the result of this particular division, that is the quotient, is a polynomial):

$$\frac{3x^2 + 5x - 2}{x + 2}$$

It is clear that the quotient must be a polynomial of degree 1: $ax + b$. In other words, it must be that:

$$3x^2 + 5x - 2 = (x + 2)(ax + b) \tag{5.15}$$

The first task is to find the coefficient a, or the first term of the unknown polynomial, ax. From Equality 5.15 we see that $ax^2 = 3x^2$. From here,

$$ax = \frac{3x^2}{x} = 3x$$

This is what is done at the beginning of the long division process (see $3x$ written above the line):

$$\begin{array}{r} 3x \\ x+2 \overline{)\ 3x^2 + 5x - 2} \end{array}$$

Thus, we know that $a = 3$. Next, we use the distributive property on the right side of Equality 5.15:

$$3x^2 + 5x - 2 = (x + 2)3x + (x + 2)b = 3x^2 + 6x + (x + 2)b$$

Or

$$3x^2 + 6x - x - 2 = 3x^2 + 6x + (x + 2)b$$

That is

$$-x - 2 = (x + 2)b$$

This allows us to see that $b = -1$. With this in mind, we finally find that $ax + b = 3x - 1$. This is what is done in the remaining part of the long division process:

$$\begin{array}{r} 3x - 1 \\ x+2 \overline{)\ 3x^2 + 5x - 2} \\ \underline{-3x^2 - 6x} \\ -x - 2 \\ \underline{x + 2} \\ 0 \end{array}$$

Hence, the long division process has essentially done the following:

$$\frac{3x^2 + 5x - 2}{x + 2} = \frac{3x^2 + 6x - x - 2}{x + 2} = \frac{3x^2 + 6x}{x + 2} - \frac{x + 2}{x + 2} = \frac{3x(x + 2)}{x + 2} - \frac{x + 2}{x + 2} = 3x - 1$$

We see that the process of long division systematically decomposes select terms of the dividend and then forms groups of terms that are divisible by the divisor. The process terminates either when all such groups have been divided, or when there is a remainder polynomial that is not divisible.

Because an indivisible polynomial may appear at the end of the process, the quotient of polynomial division is not always a polynomial. For example:

$$\frac{2x^3 - 5x^2 - 12x + 3}{x - 4} = 2x^2 + 3x + \frac{3}{x - 4} \tag{5.16}$$

This is how the process of long division is performed for 5.16:

$$
\begin{array}{r}
2x^2\ \ + 3x \\
x - 4 \overline{)\ 2x^3 - 5x^2 - 12x + 3} \\
-\ 2x^3 + 8x^2 \\
\hline
3x^2 - 12x \\
-\ 3x^2 + 12x \\
\hline
3
\end{array}
$$

Here, the dropped 3 becomes the new dividend. But is it clear that dividing 3 by $x - 4$ will not produce a polynomial, thus the process terminates.

The expression $\frac{3}{x-4}$ which is a part of the quotient in Equality 5.16 is called the *remainder*. In general, a remainder is a rational proper fraction:

$$Remainder = \frac{Polynomial\ of\ degree\ n}{Polynomial\ of\ degree\ m}$$

where $m > n$.

One more interesting operation we should discuss is polynomial *composition*. Consider two polynomials:

$$A = a_n x^n + a_{n+1} x^{n+1} + \cdots + a_1 x + a_0$$

$$B = b_m y^m + b_{m+1} y^{m+1} + \cdots + b_1 y + b_0$$

The composition $A \circ B$ (read "A of B" or "A on B") is defined in the following way:

$$A \circ B = a_n B^n + a_{n+1} B^{n+1} + \cdots + a_1 B + a_0$$

When we perform the operation $A \circ B$, we substitute polynomial B in place of the variable of polynomial A. In general, for the composition to be well-defined, polynomial B can have many variables, but polynomial A must be a one-variable polynomial.

Below are several examples of performing composition:

$$(x^2 + 3) \circ (x + 1) = (x + 1)^2 + 3 = x^2 + 2x + 4$$

$$(2x^3 - x^2) \circ y^2 = 2(y^2)^3 - (y^2)^2 = 2y^6 - y^4$$

$$x^3 \circ (2x^2 y - x) = (2x^2 y - x)^3 = 8x^6 y^3 - 12x^5 y^2 + 6x^4 y - x^3$$

The composition of polynomials always results in the creation of another polynomial.

When both A and B are one-variable polynomials, both compositions are defined: $A \circ B$ and $B \circ A$. In this case, one of the interesting characteristics of composition is that it is not a *commutative* operation. This implies that $A \circ B$ may produce a different polynomial than $B \circ A$. So in general:

$$A \circ B \neq B \circ A$$

Indeed:

$$(x^2 + 1) \circ (x - 1) = x^2 - 2x + 2 \quad but \quad (x - 1) \circ (x^2 + 1) = x^2$$

The composition is *associative*. This means that for any three polynomials A, B, and C, the order of composition of the polynomials is irrelevant:

$$A \circ (B \circ C) = (A \circ B) \circ C$$

Finally, the composition is *right-distributive*:

$$(A + B) \circ C = A \circ C + B \circ C$$

Note that the composition is not, in general, *left-distributive* (see Exercise 5.16):

$$C \circ (A + B) \neq C \circ A + C \circ B$$

We can speak about the *iterative power* for a one-variable polynomial A and a positive integer n. This power is denoted as $A^{\circ n}$ and is defined in the following way:

$$A^{\circ n} = \underbrace{A \circ A \circ A \circ \cdots \circ A}_{n \; times} \qquad (A^{\circ 1} = A)$$

For example:

$$(x^2 + 4x)^{\circ 2} = (x^2 + 4x) \circ (x^2 + 4x) = (x^2 + 4x)^2 + 4(x^2 + 4x) = x^4 + 8x^3 + 20x^2 + 16x$$

The n-th *iterative root* of polynomial A is such one-variable polynomial B (if it exists), that $B^{\circ n} = A$. We will denote the n-th iterative root as $\sqrt[\circ n]{A}$ and the square iterative root as $\sqrt[\circ]{A}$. Note that, based on the definition we have provided here, there are instances where the iterative root is not unique [4]. For example:

$$\sqrt[\circ]{9x + 4} = \begin{cases} 3x + 1 \\ -3x - 2 \end{cases}$$

This is because

$$(3x + 1) \circ (3x + 1) = 3(3x + 1) + 1 = 9x + 4$$
$$(-3x - 2) \circ (-3x - 2) = -3(-3x - 2) - 2 = 9x + 4$$

---

[4]When discussing the n-th root of a number in Chapter 2, we introduced the concept of the *principal* n-th root (and utilized the radix symbol $\sqrt{}$ exclusively for this principal root) to ensure its uniqueness. However, when it comes to the iterative root of a polynomial, we will allow it to generate multiple expressions. In the exercises of this book polynomials will have at most 2 iterative roots.

5.2   Examples of worked exercises

Example 5.1 *For polynomials $A = 2x^2 + 3x + 1$ and $B = 2x + 1$, simplify:*

$$\frac{(A - B) \circ B}{A(x + 1)^{-1}(4x + 3)}$$

Solution. 1. Simplify the numerator:

$$A - B = 2x^2 + 3x + 1 - (2x + 1) = 2x^2 + x$$

$$(A - B) \circ B = (2x^2 + x) \circ (2x + 1) = 2(2x + 1)^2 + (2x + 1) = (2x + 1)(4x + 3)$$

2. Simplify the denominator:

$$A(x + 1)^{-1}(4x + 3) = \frac{2x^2 + 3x + 1}{x + 1}(4x + 3)$$

Performing the long division

$$
\begin{array}{r}
2x + 1 \\
x + 1 \overline{)\ 2x^2 + 3x + 1} \\
-2x^2 - 2x \\
\hline
x + 1 \\
-x - 1 \\
\hline
0
\end{array}
$$

we obtain:

$$A(x + 1)^{-1}(4x + 3) = (2x + 1)(4x + 3)$$

3. Return to the original expression:

$$\frac{(A - B) \circ B}{A(x + 1)^{-1}(4x + 3)} = \frac{(2x + 1)(4x + 3)}{(2x + 1)(4x + 3)} = \boxed{1}$$

Of course, this equality is valid for all x, except $x = -1$, $x = -\frac{1}{2}$, $x = -\frac{3}{4}$.

■ **End of the example**

Example 5.2 *Find the quotient $(x \neq -2)$:*

$$\frac{x^5 - 1}{x + 2}$$

Is it a polynomial?

Solution.

$$
\begin{array}{r}
x^4 - 2x^3 + 4x^2 \quad - 8x + 16 \qquad\quad \\
\hline
x + 2 \overline{)\ x^5 \qquad\qquad\qquad\qquad\qquad -1} \\
\underline{-x^5 - 2x^4 \qquad\qquad\qquad\qquad} \\
-2x^4 \qquad\qquad\qquad\qquad \\
\underline{2x^4 + 4x^3 \qquad\qquad\qquad} \\
4x^3 \qquad\qquad\qquad \\
\underline{-4x^3 - 8x^2 \qquad\qquad} \\
-8x^2 \qquad\qquad \\
\underline{8x^2 + 16x \qquad} \\
16x \quad - 1 \\
\underline{-16x - 32} \\
-33
\end{array}
$$

The quotient is:

$$
\boxed{x^4 - 2x^3 + 4x^2 - 8x + 16 - \frac{33}{x+2}}
$$

The quotient is not a polynomial because there is a remainder of $-\frac{33}{x+2}$.

■ **End of the example**

Example 5.3 *Factor* $x^5 + 3x^3 - 4x^2 - 12$ *as a product of two binomials.*

Solution. This problem is similar to Example 4.4 and Problem 4.5. Carefully analyzing the polynomial we predict the structure to be

$$
x^5 + 3x^3 - 4x^2 - 12 = (ax^3 + b)(cx^2 + d) = acx^5 + adx^3 + bcx^2 + bd
$$

This shows that we can take $a = c = 1$, $b = -4$ and $d = 3$. Another approach to solve this problem would be to use the distributive property three times:

$$
x^5 + 3x^3 - 4x^2 - 12 = x^3(x^2 + 3) - 4(x^2 + 3) = (x^2 + 3)(x^3 - 4)
$$

Hence:

$$
x^5 + 3x^3 - 4x^2 - 12 = \boxed{(x^2 + 3)(x^3 - 4)}
$$

■ **End of the example**

Example 5.4 *Find the quotient without using long division:*

$$\frac{3x^5 + 9x^3 - 12x^2 - 36}{x^2 + 3}$$

Solution. Using our solution from Example 5.3:

$$\frac{3x^5 + 9x^3 - 12x^2 - 36}{x^2 + 3} = \frac{3(x^5 + 3x^3 - 4x^2 - 12)}{x^2 + 3} = \frac{3(x^2 + 3)(x^3 - 4)}{x^2 + 3}$$

$$= 3(x^3 - 4) = \boxed{3x^3 - 12}$$

■ **End of the example**

Example 5.5 *Complete the square for the trinomial:* $3x^2 - 12x + 16$

Solution. "Completing the square" for a trinomial $ax^2 + bx + c$ means representing it in the form:

$$a(x + b)^2 + c$$

We factor 3 out and identify the square:

$$3x^2 - 12x + 16 = 3\left(x^2 - 4x + \frac{16}{3}\right) = 3\left(x^2 - 4x + \frac{12}{3} + \frac{4}{3}\right)$$

$$= 3\left(x^2 - 4x + 4 + \frac{4}{3}\right) = 3\left((x - 2)^2 + \frac{4}{3}\right) = \boxed{3(x - 2)^2 + 4}$$

■ **End of the example**

Example 5.6 *Find the quotient without using long division* $(x \neq 2)$:

$$\frac{3x^2 - 12x + 16}{x - 2}$$

Solution. Using our solution from Example 5.5:

$$\frac{3x^2 - 12x + 16}{x - 2} = \frac{3(x - 2)^2 + 4}{x - 2} = \frac{3(x - 2)^2}{x - 2} + \frac{4}{x - 2} = \boxed{3x - 6 + \frac{4}{x - 2}}$$

The quotient is not a polynomial.

■ **End of the example**

Example 5.7 *Factor the following polynomial as a product of two binomials:*

$$(2x + 1)^{\circ 3} + \sqrt[\circ 3]{x^8}$$

Solution. We have:

$$(2x + 1)^{\circ 3} = (2x + 1) \circ (2x + 1) \circ (2x + 1)$$
$$= (2x + 1) \circ (2(2x + 1) + 1) = (2x + 1) \circ (4x + 3)$$
$$= 2(4x + 3) + 1 = 8x + 7$$

$$\sqrt[\circ 3]{x^8} = \sqrt[\circ 3]{\left((x^2)^2\right)^2} = \sqrt[\circ 3]{(x^2)^{\circ 3}} = x^2$$

Therefore:

$$(2x + 1)^{\circ 3} + \sqrt[\circ 3]{x^8} = x^2 + 8x + 7$$

Let's factor this polynomial:

$$x^2 + 8x + 7 = x^2 + x + 7x + 7 = x(x + 1) + 7(x + 1) = \boxed{(x + 7)(x + 1)}$$

■ **End of the example**

5.3   Exercises

The variables in the exercises can take negative, zero or positive numbers as long as the given expressions are well defined (in each exercise assume such variables that their values cannot result in a zero denominator).

Exercise 5.1   *Factor the following polynomials as a product of two binomials. An example of the expected answer:*

$$6x^3 + 5x^2 - 6x = (2x^2 + 3x)(3x - 2)$$

(a) $x^2 - 3x + 2$

(b) $x^2 - x - 12$

(c) $6x^2 - 7x - 5$

(d) $x^4 - 3x^2 + 2$

(e) $2x^3 - 6x^2 - x + 3$

(f) $x^6 + 3x^4 + x^2 + 3$

(g) $x^4 - x^3 - 2x^2$

(h) $x^4 - 2x^3 + 5x^2 - 10x$

(i) $x^2 + xy - 2y^2$

(j) $x^5 + 2x^3y + 2x^2 + 4y$

Exercise 5.2   *Find the quotients, without using long division. The quotients are guaranteed to be polynomials. For (a) - (h), after you have obtained the answers without using long division, divide the same expressions using long division. An example of the expected answer:*

$$\frac{4x^2 + 12x + 9}{2x + 3} = 2x + 3$$

(a) $\frac{x^4 - 16}{x + 2}$

(b) $\frac{9x^2 + 24x + 16}{3x + 4}$

(c) $\frac{9x^6 + 18x^4 + 9x^2}{x^3 + x}$

(d) $\frac{4x^6 - 8x^5 + 4x^4}{x^2 - 2x + 1}$

(e) $\frac{x^3 - 9x^2 + 27x - 27}{x - 3}$

(f) $\frac{x^3 - 8}{x - 2}$

(g) $\frac{3x^2 + 9x - 12}{x + 4}$

(h) $\frac{6x^2 + 11x + 3}{3x + 1}$

(i) $\dfrac{x^3y^6-x^3y^3}{y^2+y+1}$

(j) $\dfrac{x^3y^3+3x^2y^3+3xy^3+y^3}{xy+y}$

(k) $\dfrac{x^2+2xy+6x+y^2+6y+9}{x+y+3}$

(l) $\dfrac{x^3+3x^2y-x^2+3xy^2-2xy+y^3-y^2}{x+y-1}$

Exercise 5.3 *Complete the square for the following trinomials. An example of the expected answer:*

$$5x^2 + 10x + 3 = 5(x + 1)^2 - 2$$

(a) $3x^2 + 6x + 6$

(b) $2x^2 + 8x + 7$

(c) $-2x^2 - 16x - 37$

(d) $-2x^2 + 12x - 20$

(e) $0.5x^2 - 2x + 3$

(f) $3x^2 - 3x - 3.25$

(g) $6x^2 + 36x + 53.5$

(h) $10x^2 + 20x + 10.25$

(i) $2x^2 + 40x + 100$

(j) $100x^2 - 2000x + 9900$

Exercise 5.4 *Complete the cube for the following polynomials. An example of the expected answer:*

$$2x^3 - 12x^2 + 24x - 19 = 2(x - 2)^3 - 3$$

(a) $x^3 - 6x^2 + 12x - 9$

(b) $2x^3 + 6x^2 + 6x + 4$

(c) $-3x^3 - 18x^2 - 36x - 28$

(d) $2x^3 - 18x^2 + 54x - 44$

(e) $0.5x^3 - 1.5x^2 + 1.5x - 30.5$

(f) $4x^3 + 48x^2 + 192x + 255.5$

Exercise 5.5   *Find the quotients, without using long division. Each quotient will be a polynomial plus a non-polynomial. After you have obtained the answers without using long division, divide the same expressions using long division. An example of the expected answer:*

$$\frac{4x^2 + 12x + 8}{2x + 3} = 2x + 3 - \frac{1}{2x + 3}$$

(a) $\frac{x^2+2x}{x+1}$

(b) $\frac{x^2-8x+19}{x-4}$

(c) $\frac{-3x^2+30x-71}{x^2-10x+25}$

(d) $\frac{3x^2-18x+36}{3x-9}$

(e) $\frac{2x^3-12x^2+24x-14}{x^2-4x+4}$

(f) $\frac{4x^3+36x^2+108x+106}{2x+6}$

Hint: in (g) - (j) *look for* $a(x+b)(x+c)+d$

(g) $\frac{x^2-7}{x+2}$

(h) $\frac{2x^2-9}{x-2}$

(i) $\frac{x^2-x-3}{x+2}$

(j) $\frac{2x^2-6x+3}{x-2}$

Exercise 5.6   *Find the quotients, without using long division. The quotients may or may not be polynomials.*

(a) $\frac{9x^2-12x-11}{3x-2}$

(b) $\frac{x^3-3x^2+3x-1001}{x^2-2x+1}$

(c) $\frac{8x^3+36x^2+54x+27}{4x^2+12x+9}$

(d) $\frac{x^3-6x^2y+12xy^2-8y^3-1}{x-2y-1}$

(e) $\frac{3x^6+3x^3-12x^2}{x^4+x-4}$

(f) $\frac{x^2-4x+3}{x-3}$

(g) $\frac{x^2-3xy+2y^2+1}{x-y}$

(h) $\frac{x^2-x-13}{x-4}$

(i) $\frac{2x^3+18x^2+54x+57}{x+3}$

(j) $\frac{4x^2-y^2+1}{2x+y}$

(k) $\dfrac{x^3-12x^2y+48xy^2-64y^3}{x^2-8xy+16y^2}$

(l) $\dfrac{x^2-5xy^2-3xy+15y^3}{x-5y^2}$

(m) $\dfrac{x^3-5x^2-2x-10}{x^2-2}$

(n) $\dfrac{x^4+91}{x^2-3}$

Exercise 5.7   *Find the quotients, using long division:*

(a) $\dfrac{x^8+3x^7+2x^5+7x^4+2x}{x^3+2}$

(b) $\dfrac{x^4-6}{x-1}$

(c) $\dfrac{2x^5-5x^4+9x^3-14x^2+5x-15}{2x^3-3x^2-5}$

(d) $\dfrac{x^5-x^3+x^2-1}{x^2+1}$

(e) $\dfrac{3x^7-x^5+x^3+x^2+1}{x^3+x^2-1}$

(f) $\dfrac{x^9+x^7+x^5}{x^4+x^2}$

(g) $\dfrac{4x^4+5x^3+12x^2-2x-1}{x+4}$

(h) $\dfrac{2x^9+5x^8+8x^7+12x^6+x^5+12x^4}{x^2+x+3}$

Exercise 5.8   *Simplify the following expressions. Examples of the expected answers:*

$$(x+1)\circ(x^2+1)=x^2+2,\quad (2x+3)^{\circ 2}=4x+9,\quad \sqrt[\circ]{4x+9}=2x+3\ (and-2x-9)$$

(a) $(x+1)\circ(x-1)$

(b) $(x^2+1)\circ(x+2)$

(c) $(3x^2+2)\circ(2x-1)$

(d) $(2x-1)\circ(3x^2+2)$

(e) $(x-1)\circ(2x^3-3x^2)$

(f) $(2x^3-3x^2)\circ(x-1)$

(g) $(x^5)^{\circ 2}+(x+10)^{\circ 3}-20$

(h) $\sqrt[\circ]{x+20}+\sqrt[\circ]{x^{100}}$

Hint: each of (i) - (l) *has two iterative roots*

(i) $\sqrt[\circ]{9x-16}$

(j) $\sqrt[\circ]{25x+24}$

(k) $\sqrt[\circ]{16x+15}$

(l) $\sqrt[\circ]{4x+3}$

(m) $\sqrt[\circ]{x^4-2x^2}$

(n) $\sqrt[\circ]{x^4+2x^2+2}$

Exercise 5.9   *Simplify the following expressions. Examples of the expected answers:*

$$\frac{x^3 \circ (x-3)}{(x-3)^2} = x - 3, \quad \frac{(x^2-2)^{\circ 2}}{x^4 - 4x^2 + 2} = 1, \quad \sqrt[\circ 3]{x+9} - 3 = x$$

(a) $\dfrac{(x^2+4)\circ(x+2)-4}{x+2}$

(b) $\dfrac{(x^3-x^2)\circ(x^2-1)}{(x^2-2)(x^2\circ(x^2-1))}$

(c) $\dfrac{(x+1)^2\circ(x-2)^2}{x^2-4x+5} - x(x-1)^{\circ 4}$

(d) $\dfrac{2x^3\circ(x+1)\circ x^2}{x^4+2x^2+1} - (2x)\circ x^2$

(e) $\dfrac{x^2\circ x^3\circ(x+1)}{x^5\circ(x+1)}$

(f) $\dfrac{(x+2)^2\circ(x-3)}{(x-3)\circ(x+2)^2} \cdot \dfrac{(x^2-3)\circ(x+2)}{(x-1)\circ x}$

(g) $(x^2-2)^{\circ 2} - (x^2+2)^{\circ 2} + 8\left(x^{\circ 2}\right)^2$

(h) $\dfrac{(x-2)(x^2-4)^{\circ 2}}{(x^2-6)(x-1)^{\circ 2}}$

(i) $x^2 \circ (x+2)^{\circ 3} \circ (x+1)^{\circ 3}\left(\sqrt[\circ]{x+18}\right)^{-1}$

(j) $\sqrt[4]{\sqrt[\circ 3]{(x^8)^{\circ 2}}}$

Exercise 5.10   *Simplify the following expressions:*

(a) $\dfrac{(x^3+x^2)^{\circ 2}}{(x+1)^2(x^3+x^2+1)}$

(b) $\dfrac{(2x^3-1)\circ(2x^2+1)-12x^2-1}{2x^2+3}$

Exercise 5.11   *Simplify the expression:*

$$\frac{x^2((x^2-2)x^2+1)-3}{(x-1)^2(x+1)^2} - \frac{(x^3-3)\circ(x^2-1)}{x^4-2x^2+1}$$

Exercise 5.12   *Simplify the expression:*

$$\frac{(x+1)\circ(x+1)^4}{(x^2+2x+2)\circ(x^2+2x)}$$

Exercise 5.13   *Simplify $A/(B \circ C)$ if:*

$$A = x^4 \circ (2x-1) - x, \quad B = 16x^3 + \frac{8x}{3} + \frac{13}{27}, \quad C = x - \frac{1}{3}$$

Exercise 5.14   *Simplify $A/B - C$ if:*

$$A = (x^3 - x^2)^{\circ 2}, \quad B = x^6 - 2x^5 + x^4, \quad C = \frac{x^5 - x^2}{x^2 + x + 1}$$

Exercise 5.15   *Simplify $A^{\circ 2}/(B \circ C \circ D)$ if:*

$$A = (x^2 + 1)^{\circ 2}, \quad B = x^2 + 10x + 26, \quad C = x^2 + 4x, \quad D = x^4 + 2x^2$$

Exercise 5.16   **(a)** *Show that the composition is not, in general, left-distributive. In other words, find three polynomials A, B and C, for which:*

$$C \circ (A + B) \neq C \circ A + C \circ B$$

(b) *Show that, sometimes, the composition is left-distributive.*

(c) *Show that, sometimes, the composition is commutative. In other words, find two polynomials A and B, for which:*

$$A \circ B = B \circ A$$

Exercise 5.17   *Explain the error in the following solution and correct it.*

Denoting $k = x + 2$:

$$(x + 1) \circ (x + 2) = (k - 1) \circ k = k - 1 = x + 1$$

Exercise 5.18   *Resolve the following expressions. Hint: in (a) and (b), find the iterative powers for $n = 2, 3, 4, \ldots$, and deduct the general formula, for any positive integer n, from the emerging pattern.*

(a) $(x + 1)^{\circ n}$

(b) $(2x + 1)^{\circ n}$

(c) $\sqrt[\circ 3]{x + 45}$

(d) $\sqrt[\circ 3]{27x + 13}$

Exercise 5.19   **(a)** *For multiplication, digit 1 has the property that $A \cdot 1 = 1 \cdot A = A$. In the realm of polynomial composition, let's explore the analogue of 1. Find the polynomial E that fulfills the condition*

$$A \circ E = E \circ A = A$$

for any one-variable polynomial A.

(b) *For multiplication, $A^{-1}A = AA^{-1} = 1$. Let's define $A^{\circ -1}$ (the inverse of the one-variable polynomial A) as a one-variable polynomial (if it exists), for which*

$$A^{\circ -1} \circ A = A \circ A^{\circ -1} = E$$

where E is found in part (a). Find $(2x + 1)^{\circ -1}$.

(c) *Let's define (m is an integer[5], n is a positive integer):*

$$A^{\circ \frac{m}{n}} = \left(\sqrt[\circ n]{A} \right)^{\circ m}$$

Using this definition, simplify:

$$(8x + 7)^{\circ \frac{2}{3}} + (27x + 13)^{\circ -\frac{1}{3}}$$

Exercise 5.20   *For any integer $n > 0$, find:*

(a) $\frac{x^n - 1}{x - 1}$

(b) $\frac{x^n + 1}{x + 1}$

   Direction: for both (a) and (b), divide for $n = 2, 3, 4, \ldots$, and deduct the general formula from the emerging pattern. Then, try to resolve (a) without resorting to long division.

---

[5]We already know how to define $P^{\circ -1}$, for some polynomial P. While it is not necessary for this exercise, you might wonder what to do if m is a negative number smaller than -1. For example, for $m = -2$, we can define: $P^{\circ -2} = (P^{\circ 2})^{\circ -1}$.

5.4 Answers

5.1: **(a)** $(x-1)(x-2)$ **(b)** $(x+3)(x-4)$ **(c)** $(2x+1)(3x-5)$ **(d)** $(x^2-2)(x^2-1)$ **(e)** $(2x^2-1)(x-3)$ **(f)** $(x^2+3)(x^4+1)$ **(g)** $(x^2+x)(x^2-2x)$ **(h)** $(x^2-2x)(x^2+5)$ **(i)** $(x+2y)(x-y)$ **(j)** $(x^2+2y)(x^3+2)$

5.2: **(a)** x^3-2x^2+4x-8 **(b)** $3x+4$ **(c)** $9x^3+9x$ **(d)** $4x^4$ **(e)** x^2-6x+9 **(f)** x^2+2x+4 **(g)** $3x-3$ **(h)** $2x+3$ **(i)** $x^3y^4-x^3y^3$ **(j)** $x^2y^2+2xy^2+y^2$ **(k)** $x+y+3$ **(l)** $x^2+2xy+y^2$

5.3: **(a)** $3(x+1)^2+3$ **(b)** $2(x+2)^2-1$ **(c)** $-2(x+4)^2-5$ **(d)** $-2(x-3)^2-2$ **(e)** $0.5(x-2)^2+1$ **(f)** $3(x-0.5)^2-4$ **(g)** $6(x+3)^2-0.5$ **(h)** $10(x+1)^2+0.25$ **(i)** $2(x+10)^2-100$ **(j)** $100(x-10)^2-100$

5.4: **(a)** $(x-2)^3-1$ **(b)** $2(x+1)^3+2$ **(c)** $-3(x+2)^3-4$ **(d)** $2(x-3)^3+10$ **(e)** $0.5(x-1)^3-30$ **(f)** $4(x+4)^3-0.5$

5.5: **(a)** $x+1-\frac{1}{x+1}$ **(b)** $x-4+\frac{3}{x-4}$ **(c)** $-3+\frac{4}{(x-5)^2}$ **(d)** $x-3+\frac{3}{x-3}$ **(e)** $2x-4+\frac{2}{(x-2)^2}$ **(f)** $2x^2+12x+18-\frac{1}{x+3}$ **(g)** $x-2-\frac{3}{x+2}$ **(h)** $2x+4-\frac{1}{x-2}$ **(i)** $x-3+\frac{3}{x+2}$ **(j)** $2x-2-\frac{1}{x-2}$

5.6: **(a)** $3x-2-\frac{15}{3x-2}$ **(b)** $x-1-\frac{1000}{(x-1)^2}$ **(c)** $3+2x$ **(d)** $x^2+4y^2-4xy+x-2y+1$ **(e)** $3x^2$ **(f)** $x-1$ **(g)** $x-2y+\frac{1}{x-y}$ **(h)** $x+3-\frac{1}{x-4}$ **(i)** $2x^2+12x+18+\frac{3}{x+3}$ **(j)** $2x-y+\frac{1}{2x+y}$ **(k)** $x-4y$ **(l)** $x-3y$ **(m)** $x-5-\frac{20}{x^2-2}$ **(n)** $x^2+3+\frac{100}{x^2-3}$

5.7: **(a)** x^5+3x^4+x **(b)** $x^3+x^2+x+1-\frac{5}{x-1}$ **(c)** x^2-x+3 **(d)** $x^3-2x+1+\frac{2x-2}{x^2+1}$ **(e)** $3x^4-3x^3+2x^2+x-3+\frac{6x^2+x-2}{x^3+x^2-1}$ **(f)** $x^5+x-\frac{x}{x^2+1}$ **(g)** $4x^3-11x^2+56x-226+\frac{903}{x+4}$ **(h)** $2x^7+3x^6-x^5+4x^4$

5.8: **(a)** x **(b)** x^2+4x+5 **(c)** $12x^2-12x+5$ **(d)** $6x^2+3$ **(e)** $2x^3-3x^2-1$ **(f)** $2x^3-9x^2+12x-5$ **(g)** $x^{25}+x+10$ **(h)** $x^{10}+x+10$ **(i)** $3x-4$ and $-3x+8$ **(j)** $5x+4$ and $-5x-6$ **(k)** $4x+3$ and $-4x-5$ **(l)** $2x+1$ and $-2x-3$ **(m)** x^2-1 **(n)** x^2+1

5.9: **(a)** $x+2$ **(b)** 1 **(c)** 5 **(d)** 2 **(e)** $x+1$ **(f)** $x-1$ **(g)** -4 **(h)** x^2-2 **(i)** $x+9$ **(j)** $|x|$

5.10: **(a)** x^4 **(b)** $8x^4$

5.11: 1

5.12: 1

5.13: $x-1$

5.14: -1

5.15: 1

5.18: **(a)** $x+n$ **(b)** 2^nx+2^n-1 **(c)** $x+15$ **(d)** $3x+1$

The answers to 5.19 and 5.20 are on the next page.

5.19: **(a)** $E = x$ **(b)** $0.5x - 0.5$ **(c)** $\frac{13}{3}x + \frac{8}{3}$

5.20:

(a)

$$x^{n-1} + x^{n-2} + x^{n-3} + \cdots + x + 1$$

(b) For odd n:

$$(-x)^{n-1} + (-x)^{n-2} + (-x)^{n-3} + \cdots + (-x)^1 + 1$$

and for even n:

$$(-1)^n x^{n-1} + (-1)^{n-1} x^{n-2} + (-1)^{n-2} x^{n-3} + \cdots + (-1)^2 x^1 - 1 + \frac{2}{x - 1}$$

Chapter 6

Factorial, sigma, and product notations

6.1   Explanation

There are instances when long algebraic expressions can be compactly written using special notations. We will discuss several such instances.

You have already encountered[1] the *factorial* notation "$n!$" (read "n factorial") that represents the product of all positive integers up to and including a positive integer n:

$$n! = 1 \cdot 2 \cdot 3 \cdot \; \cdots \; \cdot n \tag{6.1}$$

For example:

$$5! = 1 \cdot 2 \cdot 3 \cdot 4 \cdot 5 = 120$$

The factorial of 0 is *defined* to be equal to 1 (that is $0! = 1$).

The *double factorial* $n!!$ (read "n double factorial") of a positive integer n is defined in the following way:

$$n!! = \begin{cases} 1 \cdot 3 \cdot 5 \cdot \; \cdots \; \cdot n, & \text{if } n \text{ is odd} \\ 2 \cdot 4 \cdot 6 \cdot \; \cdots \; \cdot n, & \text{if } n \text{ is even} \end{cases} \tag{6.2}$$

For example:

$$5!! = 1 \cdot 3 \cdot 5 = 15$$

$$6!! = 2 \cdot 4 \cdot 6 = 48$$

---

[1]See Example 2.5 and Exercise 2.18.

The double factorial of 0 is *defined* to be equal to 1 (that is $0!! = 1$).

The so-called *sigma notation* helps in writing sums of many terms compactly. This notation uses the capital Greek letter Σ (sigma).

Given n numbers or algebraic expressions x_1, x_2, ... , x_n, the sigma notation is defined in the following way:

$$\sum_{i=1}^{n} x_i = x_1 + x_2 + x_3 + \cdots + x_n \tag{6.3}$$

The left side of Equality 6.3 is read "sum of x_i, from $i = 1$ to n".

For example, for $x_1 = 2$, $x_2 = 5$, $x_3 = 10$, $x_4 = -4$, $x_5 = 3$, we have:

$$\sum_{i=1}^{5} x_i = x_1 + x_2 + x_3 + x_4 + x_5 = 2 + 5 + 10 - 4 + 3 = 16$$

The index i alone may be enough to write many sums compactly. Consider the following examples:

$$\sum_{i=1}^{5} i = 1 + 2 + 3 + 4 + 5 = 15$$

$$\sum_{i=1}^{6} i^2 = 1^2 + 2^2 + 3^2 + 4^2 + 5^2 + 6^2 = 91$$

$$\sum_{i=1}^{7} (-1)^i i = -1 + 2 - 3 + 4 - 5 + 6 - 7 = -4$$

We do not have to start with the index $i = 1$. For integers m and n, such that $n \geq m$, we have:

$$\sum_{i=m}^{n} x_i = x_m + x_{m+1} + x_{m+2} + \cdots + x_n \tag{6.4}$$

For example:

$$\sum_{i=-6}^{-4} (i^3 - i^2) = ((-6)^3 - (-6)^2) + ((-5)^3 - (-5)^2) + ((-4)^3 - (-4)^2) = -482$$

The sigma notation can be combined; that is, we can use several sigmas to form sums of sums[2]:

$$\sum_{i} \sum_{j} x_{ij} \qquad \sum_{i} \sum_{j} \sum_{k} x_{ijk}$$

---

[2]Notice that index specifications are omitted from the notation when it is evident from the context what they represent.

Below are examples of combining two sigmas:

$$\sum_{i=1}^{4}\sum_{j=2}^{3} i^2 j^3 = \sum_{i=1}^{4}\left(\sum_{j=2}^{3} i^2 j^3\right) = \sum_{i=1}^{4}(i^2 \cdot 2^3 + i^2 \cdot 3^3)$$

$$= (1^2 \cdot 2^3 + 1^2 \cdot 3^3) + (2^2 \cdot 2^3 + 2^2 \cdot 3^3) + (3^2 \cdot 2^3 + 3^2 \cdot 3^3) + (4^2 \cdot 2^3 + 4^2 \cdot 3^3) = 1050$$

$$\sum_{j=2}^{3}\sum_{i=1}^{4} i^2 j^3 = \sum_{j=2}^{3}\left(\sum_{i=1}^{4} i^2 j^3\right) = \sum_{j=2}^{3}(1^2 \cdot j^3 + 2^2 \cdot j^3 + 3^2 \cdot j^3 + 4^2 \cdot j^3)$$

$$= (1^2 \cdot 2^3 + 2^2 \cdot 2^3 + 3^2 \cdot 2^3 + 4^2 \cdot 2^3) + (1^2 \cdot 3^3 + 2^2 \cdot 3^3 + 3^2 \cdot 3^3 + 4^2 \cdot 3^3) = 1050$$

The above examples also demonstrate that sigma notation is commutative. That is, we can change the order of sigmas[3]:

$$\sum_i \sum_j x_{ij} = \sum_j \sum_i x_{ij} \tag{6.5}$$

Here are several other useful properties of the sigma notation:

$$\sum_i a x_i = a \left(\sum_i x_i\right) \tag{6.6}$$

$$\sum_i (x_i + y_i) = \sum_i x_i + \sum_i y_i \tag{6.7}$$

$$\sum_i \sum_j x_i y_j = \left(\sum_i x_i\right)\left(\sum_j y_j\right) \tag{6.8}$$

It is not hard to see why Equalities 6.6 and 6.7 are valid. To demonstrate the validity of Equality 6.8, we will consider the case where $i = 1, 2, 3$ and $j = 1, 2$. The validity of the general case will become transparent from this demonstration.

$$\sum_{i=1}^{3}\sum_{j=1}^{2} x_i y_j = \sum_{i=1}^{3}(x_i y_1 + x_i y_2)$$

$$= (x_1 y_1 + x_1 y_2) + (x_2 y_1 + x_2 y_2) + (x_3 y_1 + x_3 y_2)$$

$$= x_1(y_1 + y_2) + x_2(y_1 + y_2) + x_3(y_1 + y_2)$$

$$= (x_1 + x_2 + x_3)(y_1 + y_2) = \sum_{i=1}^{3} x_i \sum_{j=1}^{2} y_j$$

---

[3]Of course, in Equality 6.5, i and j run through the same values on the left side as they do on the right side.

Returning to our earlier example of combining two sigmas, we see now that a simpler solution would be possible thanks to Equality 6.8:

$$\sum_{i=1}^{4}\sum_{j=2}^{3} i^2 j^3 = \sum_{i=1}^{4} i^2 \sum_{j=2}^{3} j^3 = (1^2 + 2^2 + 3^2 + 4^2)(2^3 + 3^3) = 1050$$

We can use the sigma notation to compactly write several useful (and elegant) formulas that you already obtained[4]:

$$\sum_{i=1}^{n} i = \frac{n(n+1)}{2} \tag{6.9}$$

$$\sum_{i=1}^{n} i^2 = \frac{n(n+1)(2n+1)}{6} \tag{6.10}$$

$$\sum_{i=1}^{n} i^3 = \left(\sum_{i=1}^{n} i\right)^2 = \frac{n^2(n+1)^2}{4} \tag{6.11}$$

$$\sum_{i=0}^{n-1} x^i = \frac{1 - x^n}{1 - x} \tag{6.12}$$

Similarly to the sigma notation, the *product notation* helps in writing complex expressions with repeated multiplication compactly. The product notation uses the capital Greek letter Π (pi). The definition of the product notation is similar to the definition of the sigma notation 6.4: for integers m and n such that $n \geq m$, we have:

$$\prod_{i=m}^{n} x_i = x_m \cdot x_{m+1} \cdot x_{m+2} \cdot \;\cdots\; \cdot x_n \tag{6.13}$$

For example, if $x_1 = -2$, $x_2 = 5$, $x_3 = -1$, $x_4 = 3$:

$$\prod_{i=1}^{4} x_i = (-2) \cdot 5 \cdot (-1) \cdot 3 = 30$$

Of course, the factorial can be written using the product notation:

$$\prod_{i=1}^{n} i = n!$$

The following example follows from Formula 6.9:

$$\prod_{i=1}^{n} x^i = x^{\frac{n^2+n}{2}}$$

---

[4]See Exercises 3.20, 4.22, 4.23 and 5.20.

The product notation can be combined:

$$\prod_{i=2}^{4}\prod_{j=1}^{2}\left(i+(-1)^j\right) = \prod_{i=2}^{4}(i-1)(i+1) = \prod_{i=2}^{4}(i^2-1) = (2^2-1)(3^2-1)(4^2-1) = 360$$

Let's mention several useful (and simply obtainable) properties of the product notation:

$$\prod_{i=1}^{n} ax_i = a^n \prod_{i=1}^{n} x_i \tag{6.14}$$

$$\left(\prod_i x_i\right)^a = \prod_i x_i^a \tag{6.15}$$

$$\prod_i \prod_j x_i y_j = \left(\prod_i x_i\right)\left(\prod_j y_j\right) \tag{6.16}$$

$$\prod_i \prod_j x_{ij} = \prod_j \prod_i x_{ij} \tag{6.17}$$

6.2   Examples of worked exercises

Example 6.1 *Calculate for $x_1 = 10, x_2 = 5, x_3 = -4, x_4 = -1$:*

$$\frac{(-x_3)! \sum_{i=2}^{4}(x_i^3 - x_i^2)}{\prod_{i=1}^{3}(x_i - 2)}$$

Solution. Using the definitions of factorial, sigma, and product notations (see 6.1, 6.4 and 6.13), we obtain:

$$\frac{(-x_3)! \sum_{i=2}^{4}(x_i^3 - x_i^2)}{\prod_{i=1}^{3}(x_i - 2)}$$
$$= 4! \cdot \frac{(5^3 - 5^2) + ((-4)^3 - (-4)^2) + ((-1)^3 - (-1)^2)}{(10 - 2)(5 - 2)(-4 - 2)} = 1 \cdot 2 \cdot 3 \cdot 4 \cdot \left(-\frac{1}{8}\right) = \boxed{-3}$$

■ **End of the example**

Example 6.2 *Simplify $(n \neq 1)$:*

$$\frac{\prod_{i=-1}^{2} \sqrt{n}(ni + 1)}{\sum_{i=1}^{n} \sum_{j=1}^{2n}(1 - n)ij}$$

Solution. 1. Work with the numerator (use Formula 6.14 before opening the product):

$$\prod_{i=-1}^{2} \sqrt{n}(ni + 1) = \left(\sqrt{n}\right)^4 \prod_{i=-1}^{2}(ni + 1)$$
$$= n^2(n \cdot (-1) + 1)(n \cdot 0 + 1)(n \cdot 1 + 1)(n \cdot 2 + 1)$$
$$= n^2(1 - n)(n + 1)(2n + 1)$$

2. Work with the denominator (use Formulas 6.6, 6.8 and 6.9):

$$\sum_{i=1}^{n} \sum_{j=1}^{2n}(1 - n)ij = (1 - n) \sum_{i=1}^{n} \sum_{j=1}^{2n} ij$$
$$= (1 - n)\left(\sum_{i=1}^{n} i\right)\left(\sum_{j=1}^{2n} j\right) = (1 - n) \cdot \frac{n(n + 1)}{2} \cdot \frac{2n(2n + 1)}{2}$$
$$= \frac{n^2(1 - n)(n + 1)(2n + 1)}{2}$$

3. Return to the original expression:

$$\frac{\prod_{i=-1}^{2} \sqrt{n}(ni+1)}{\sum_{i=1}^{n} \sum_{j=1}^{2n}(1-n)ij} = \frac{2n^2(1-n)(n+1)(2n+1)}{n^2(1-n)(n+1)(2n+1)} = \boxed{2}$$

■ **End of the example**

Example 6.3 *Simplify:*

$$\frac{(2n)!!(2n+1)!!}{(2n-1)!} \left(\sum_{i=1}^{2n} i \right)^{-1}$$

Solution. 1. Work with the fraction using Formulas 6.1 and 6.2:

$$\frac{(2n)!!(2n+1)!!}{(2n-1)!} = \frac{(2 \cdot 4 \cdot 6 \cdot \cdots \cdot 2n)(3 \cdot 5 \cdot 7 \cdot \cdots \cdot (2n+1))}{2 \cdot 3 \cdot 4 \cdot \cdots \cdot (2n-1)}$$

$$= \frac{2 \cdot 3 \cdot 4 \cdot \cdots \cdot (2n-1)(2n)(2n+1)}{2 \cdot 3 \cdot 4 \cdot \cdots \cdot (2n-1)} = 2n(2n+1)$$

2. Work with the factor containing the sum using Formula 6.9:

$$\left(\sum_{i=1}^{2n} i \right)^{-1} = \left(\frac{2n(2n+1)}{2} \right)^{-1} = \frac{1}{n(2n+1)}$$

3. Return to the original expression:

$$\frac{(2n)!!(2n+1)!!}{(2n-1)!} \left(\sum_{i=1}^{2n} i \right)^{-1} = \frac{2n(2n+1)}{n(2n+1)} = \boxed{2}$$

■ **End of the example**

6.3   Exercises

The variables in the exercises can take negative, zero, or positive numbers as long as the given expressions are well defined in the context of real-valued algebra (in each exercise assume such variables that their values cannot result in a zero denominator, an even root of a negative number or a factorial of a negative or fractional number).

Exercise 6.1   *Rewrite the following expressions, using the sigma or product notation. Examples of the expected answers:*

$$\frac{1}{9} + \frac{1}{4} + 1 + 6 + 49 + \cdots + 10^5 = \sum_{i=3}^{10} i^{i-5}$$

$$(-9)(16)(-25) \cdots 100 = \prod_{i=3}^{10} (-1)^i i^2$$

(a) $x_1^2 + x_2^2 + \cdots + x_n^2$

(b) $x_1 y_2 + x_2 y_3 + \cdots + x_{n-1} y_n$

(c) $4\sqrt{A_5} + 5\sqrt{A_6} + \cdots + 99\sqrt{A_{100}}$

(d) $(1.02\gamma_5)(1.03\gamma_6) \cdots (1.77\gamma_{80})$

(e) $\frac{z^2}{3} + \frac{z^3}{4} + \cdots + \frac{z^{23}}{24}$

(f) $\frac{-2-1+1+2+\cdots+k}{(k-1)^2 (k-2)^3 (k-3)^4 \cdots 1}$

(g) $-2 + 4 - 6 + 8 - \cdots + 200$

(h) $(-2)(3)(-4)(5) \cdots (99)$

(i) $729 - 512 + 343 - \cdots - 343 + 512 - 729$

(j) $(-0.5)(1.5)(-2.5)(3.5) \cdots (-30.5)$

(k) $a_1^{1+2+\cdots+n} + a_2^{1+2+\cdots+n} + \cdots + a_m^{1+2+\cdots+n}$

(l) $1 \cdot 3 \cdot 6 \cdot 10 \cdots 20100$

Exercise 6.2   *Evaluate the following expressions. An example of the expected answer:*

$$\sum_{k=-5}^{-2} (-1)^k (k - k^2) = 16$$

(a) $\displaystyle\sum_{k=1}^{4} \frac{W_k}{k^2 + 1}$ $W_1 = 4$ $W_2 = -5$ $W_3 = 50$ $W_4 = -34$

(b) $\displaystyle\prod_{s=1}^{4}(-1)^{s+1}\sqrt{s+r_s}$ $\qquad r_1=3, \quad r_2=-1, \quad r_3=6, \quad r_4=12$

(c) $\displaystyle\sum_{i=1}^{5}(-1)^i x_i^{i-3}$ $\qquad x_1=0.25, \quad x_2=-0.5, \quad x_3=987, \quad x_4=-3, \quad x_5=10$

(d) $\displaystyle\prod_{j=1}^{5}(-1)^j (j\phi_j)^{j-5}$ $\qquad \phi_1=\dfrac{1}{2}, \quad \phi_2=\dfrac{1}{6}, \quad \phi_3=\dfrac{1}{12}, \quad \phi_4=1728, \quad \phi_5=646$

(e) $\displaystyle\sum_{r=1}^{4} r^r - \sum_{r=-2}^{3}(-1)^r r^3$

(f) $\displaystyle 2^{-4}\left(\prod_{j=-4}^{-2}(2j+2) - \prod_{j=2}^{4}2\left(j+(-1)^{j+1}\right)^{j-1}\right)$

(g) $\displaystyle\sum_{k=25}^{50} k - \sum_{k=10}^{15}k^2$

Exercise 6.3 *Evaluate the following expressions:*

(a) $\sum_{j=1}^{4}\sum_{i=1}^{j}(-1)^j i^2$

(b) $\sum_{k=-2}^{2}\prod_{s=0}^{2}(k+(-1)^{ks}s)$

(c) $\prod_{j=1}^{2}\prod_{i=0}^{j}(1+ij)(-1)^{1+ij}$

(d) $6^{-3}\prod_{k=-3}^{-1}\sum_{s=-2}^{0}2ks$

Exercise 6.4 *Simplify the following expressions:*

(a) $\dfrac{n!}{(n-1)!}+\dfrac{(2n)!!}{(2n-2)!!}$

(b) $\dfrac{(2n+1)!!}{(2n-1)!!}-\dfrac{(2n)!}{(2n-1)!}$

(c) $\dfrac{(4!+6)n!28!!}{(n-1)!30!!}$

(d) $\dfrac{87!!(n+2)!}{(4!+5!!+6!!)(n+1)!85!!}$

Exercise 6.5 *Simplify the following expressions:*

(a) $\displaystyle\sum_{n=1}^{k}\dfrac{(2n)!!(2n+1)!!}{(k+2)(2n)!}$

(b) $\dfrac{1}{(s-1)!}\left(\displaystyle\sum_{x=s-1}^{s}\dfrac{(x!)!}{(x!-1)!}\right)$

(c) $\displaystyle\sum_{d=0}^{2}\left(\dfrac{(2d^2)!!}{2(2d^2-2)!!}\right)!$

(d) $\left(\dfrac{1}{2}\displaystyle\prod_{n=1}^{2}\dfrac{(n^2+1)!}{2(n^2-1)!}\right)!!$

(e) $\dfrac{(5!!)!!}{17!!}\displaystyle\sum_{m=0}^{2}\left(xm!+\sum_{n=0}^{2}\dfrac{x(mn)!}{m!n!}\right)$

Exercise 6.6   *Simplify the expression:*

$$\frac{1}{k!}\prod_{n=1}^{k}\frac{1}{(n-1)!}\sqrt[4]{\frac{((n!)^2+n!)^3-(n!-(n!)^2)^3}{2((n!)^2+3)}}$$

Exercise 6.7   *Simplify the expression:*

$$\frac{(n-2)!}{(n+2)!}\sum_{k=1}^{n}A_kB_k$$

where

$$A_k=\left(\sqrt{2k^{-1}}-\sqrt{k}\right)^2\left(\sqrt{2}+k\right)^2-k^3\quad and\quad B_k=\sum_{i=1}^{k}(2i-1)$$

Exercise 6.8   *Simplify the expression:*

$$\frac{1}{72n^2+18n+1}\sum_{s=1}^{6n}A_s$$

where

$$A_s=s-\frac{4}{s^2+s}\left(1+\sum_{k=2}^{s}B_k\right)\quad and\quad B_k=4k^2\left(\frac{1-k}{1+k}-\frac{1+k}{1-k}\right)^{-1}+k$$

Exercise 6.9 *Simplify the expression:*

$$1 + \frac{3}{2^n - 2} \sum_{k=0}^{n-1} (4^k - 2^k)$$

Exercise 6.10 *Simplify the expression:*

$$\sqrt[n]{1 + \frac{(x-1)}{x} \sum_{k=1}^{n} \left(\frac{2^{-x-1}(2x)!}{x!(2x-3)!!} + \frac{1}{2} \right)^k}$$

Exercise 6.11 *Simplify the expression:*

$$\sum_{n=2}^{10} A_n B_n C_n$$

where

$$A_n = \frac{x^4 - x^{n+3}}{x^{n+2} - 1}, \qquad B_n = \sum_{k=-1}^{n} 2^{n+1} x^k, \qquad C_n = \left(\sum_{s=3}^{n+1} 2x^s \right)^{-1}$$

Exercise 6.12 **(a)** *For $d \neq 0$, the sequence of n numbers*

$$a, a+d, a+2d, a+3d, \ldots, a+(n-1)d$$

*is known as an **arithmetic progression** with the common difference d. Obtain the formula for the sum of the arithmetic progression[5]:*

$$\sum_{i=0}^{n-1} (a + id) = ?$$

(b) *If a_1 is the first term of the arithmetic progression and a_n is the last term, express the sum of the arithmetic progression through n, a_1 and a_n.*

Using the formulas you found, answer (c) - (e):

(c) *Find the sum of the following 51 numbers:*

$$3, 5, 7, 9, \ldots, 103$$

---

[5]The sum of an arithmetic progression is also known as the *arithmetic series*.

(d) *Find the sum of the following 23 numbers:*

$$108, 103, 98, 93, \ldots, -2$$

(e) *Find the sum:*

$$\sum_{i=2}^{20} (3i - 100)$$

Exercise 6.13 **(a)** *For $g \neq 0, r \neq 0$ and $r \neq 1$, the sequence of n numbers*

$$g, \, gr, \, gr^2, \, gr^3, \ldots, \, gr^{n-1}$$

*is known as a **geometric progression** with the common ratio r. Obtain the formula for the sum of the geometric progression[6]:*

$$\sum_{i=0}^{n-1} gr^i = ?$$

(b) *If g_1 is the first term of the geometric progression and g_n is the last term, express the sum of the geometric progression through the common ratio r, g_1 and g_n.*

   Using the formulas you found, answer (c) *-* (e):

(c) *Find the sum of the following 9 numbers:*

$$2, 4, 8, 16, \ldots, 512$$

(d) *Find the sum of the following 8 numbers:*

$$3, 6, 12, 24, \ldots, 384$$

(e) *Find the sum:*

$$\sum_{i=1}^{10} \frac{(-2)^i}{2}$$

---

[6]The sum of a geometric progression is also known as the *geometric series*.

Exercise 6.14   *Consider the following expression:*

$$\mu = \frac{\sum_{i=1}^{n} x_i}{n}$$

μ *is known as the* **arithmetic mean** *or* **arithmetic average** *of numbers* $x_1, x_2, x_3, \ldots x_n$.
(a) *Find the arithmetic mean of the following sequence:*

$$n, \ -1, \ 2^2 n, \ -2^3, \ 3^2 n, \ -3^3, \ \ldots, \ n^3, \ -n^3$$

(b) *Find the arithmetic mean of the sequence from part* (a), *if* $n = 24$.
(c) *Find the arithmetic mean of the following sequence:*

$$-50, -46, -42, -38, \ldots, 346$$

(d) *If* a_1 *is the first term of an arithmetic progression and* a_n *is the last term, express the arithmetic mean of the arithmetic progression through* a_1 *and* a_n.
(e) *Show that for any three terms of an arithmetic progression,* a_{i-1}, a_i *and* a_{i+1}, *term* a_i *is the arithmetic average of terms* a_{i-1} *and* a_{i+1} $(i = 2, 3, \ldots, n - 1)$.

Exercise 6.15   *Consider the following expression:*

$$\sigma^2 = \frac{\sum_{i=1}^{n}(x_i - \mu)^2}{n}$$

σ^2 *is known as the* **variance**[7] *of a "population". Here the population is simply a sequence of numbers* $x_1, x_2, x_3, \ldots x_n$.
(a) *Find the variance of the following population:*

$$-10, -9, -8, -7, \ldots, 12$$

(b) *If* ν *denotes the arithmetic mean of* $x_1^2, x_2^2, x_3^2, \ldots x_n^2$, *show that:*

$$\sigma^2 = \nu - \mu^2$$

---

[7]The variance is a measure that shows the extent to which the numbers in a population deviate from their arithmetic mean. Specifically, it represents the average quadratic deviation of these numbers from their mean. Expressed in units squared, the variance is denoted as σ^2. The square root of the variance is referred to as the *standard deviation* of the population and is denoted as σ: $\sqrt{\sigma^2} = \sigma$. Unlike the variance, the standard deviation is presented in terms of the original units of the data, making it very useful in applications.

Exercise 6.16   *Consider the following expression:*

$$\gamma = \sqrt[n]{\prod_{i=1}^{n} x_i}$$

γ *is known as the **geometric mean** or **geometric average** of non-negative numbers* $x_1, x_2, x_3, \ldots x_n$.

(a) *Find the geometric mean of the following numbers:* $1, 2^n, 3^n, \ldots, n^n$.

(b) *Find the geometric mean of the following geometric progression:*

$$2, 2^2, 2^3, \ldots, 2^n$$

(c) *Find the geometric mean of the following geometric progression:*

$$3, 3 \cdot 2, 3 \cdot 2^2, 3 \cdot 2^3, \ldots, 3 \cdot 2^{n-1}$$

(d) *Find the geometric mean of the following geometric progression* $(g, r \geq 0)$:

$$g, gr, gr^2, gr^3, \ldots, gr^{n-1}$$

(e) *Given a geometric progression with non-negative terms, express its geometric mean through the first term,* g_1, *and the last term,* g_n.

(f) *Show that in a geometric progression with non-negative terms, term* g_i *is the geometric mean of terms* g_{i-1} *and* g_{i+1} $(i = 2, 3, \ldots, n-1)$.

(g) *Estimate the following expression (hint:* $1.02^{70} = 3.9996$):

$$\sqrt[71]{\prod_{i=0}^{70} 1.02^i}$$

Exercise 6.17   *Consider the following expression:*

$$d = \sqrt{\sum_{i=1}^{n} (x_i - y_i)^2}$$

*d is known as the **Euclidean distance** between two sequences:* $x_1, x_2, x_3, \ldots, x_n$ *and* $y_1, y_2, y_3, \ldots, y_n$. *We say that a sequence has size n, if it contains n numbers*[8]. *Find the squared Euclidean distance between two sequences of the same size:* $1, 2, 3, \ldots, 12$, *and* $1, 1, 1, \ldots, 1$.

---

[8]The Euclidean distance for sequences of size 3 represents the distance in the three-dimensional world we inhabit.

Exercise 6.18 *For non-negative integers k and n, $k \leq n$, consider the following expression:*

$$P_k^n = \frac{n!}{(n-k)!}$$

P_k^n *is known as the **permutation of** k **from** n. This is the number of ways k objects can be chosen from n distinct objects, without repetition, such that the order of choice matters.*

(a) *In how many ways can 5 people be chosen from a group of 7 people to be seated in a 5-seat car, given that the exact seat each person will occupy is important?*

(b) *In how many ways can the order of n objects be rearranged?*

(c) *Five friends race each other in a 100-meter sprint. In how many ways can they finish the race?*

Exercise 6.19 *For non-negative integers k and n, $k \leq n$, consider the following expression:*

$$C_k^n = \frac{n!}{k!(n-k)!}$$

C_k^n *is known as the **combination of** k **from** n. This is the number of ways k objects can be chosen from n distinct objects, without repetition, such that the order of choice does not matter.*

(a) *In how many ways can 5 people be chosen from a group of 7 people to be seated in a 5-seat car, so that the exact seat of each person is not important?*

(b) *Show that*

$$C_{n-k}^n = C_k^n$$

(c) *Show that*

$$C_{k-1}^{n-1} + C_k^{n-1} = C_k^n$$

Exercise 6.20   *The formula known as the **binomial expansion** or **binomial theorem** is the following:*

$$(a+b)^n = \sum_{k=0}^{n} C_k^n a^{n-k} b^k$$

*where a and b are any numbers, n is a non-negative integer, and C_k^n are the combinations of k from n. Note that in the context of the binomial expansion, these combinations are called **binomial coefficients**.*

(a) *Use the binomial theorem to obtain the expansion formulas for $(a+b)^2$ and $(a+b)^3$, that you are already familiar with.*

(b) *Obtain the expansion formula for $(a+b)^4$.*

(c) *Obtain the expansion formula for $(a+b)^5$.*

(d) *Using the binomial theorem, show that the following equality holds:*

$$\sum_{k=0}^{n} C_k^n = 2^n$$

(e) *The binomial coefficients form the so-called **Pascal's triangle**, shown below for the values of n up to 5:*

$$
\begin{array}{c}
n=0 \qquad\qquad\qquad 1 \\
n=1 \qquad\qquad\quad 1 \quad 1 \\
n=2 \qquad\qquad 1 \quad 2 \quad 1 \\
n=3 \qquad\quad 1 \quad 3 \quad 3 \quad 1 \\
n=4 \qquad 1 \quad 4 \quad 6 \quad 4 \quad 1 \\
n=5 \quad 1 \quad 5 \quad 10 \quad 10 \quad 5 \quad 1
\end{array}
$$

Pascal's Triangle is a triangular arrangement of numbers, where each row contains elements that are the sums of the corresponding elements in the row directly above it. The first and last elements in each row are always 1.

By continuing this triangle for $n = 6$, write the expansion formula for $(a+b)^6$.

6.4 Answers

6.1:

(a) $\sum_{i=1}^{n} x_i^2$ **(b)** $\sum_{i=1}^{n-1} x_i y_{i+1}$ or $\sum_{i=2}^{n} x_{i-1} y_i$

(c) $\sum_{i=5}^{100} (i-1)\sqrt{A_i}$ or $\sum_{i=4}^{99} i\sqrt{A_{i+1}}$ **(d)** $\prod_{i=5}^{80} \left(1 + \frac{i-3}{100}\right)\gamma_i$ or $\prod_{i=2}^{77} \left(1 + \frac{i}{100}\right)\gamma_{i+3}$

(e) $\sum_{i=2}^{23} \frac{z^i}{i+1}$ or $\sum_{i=3}^{24} \frac{z^{i-1}}{i}$ **(f)** $\frac{\sum_{i=-2}^{k} i}{\prod_{i=1}^{k-1}(k-i)^{i+1}}$ **(g)** $2\sum_{i=1}^{100}(-1)^i i$ **(h)** $\prod_{i=2}^{99}(-1)^{i+1} i$

(i) $\sum_{i=-9}^{9}(-1)^i i^3 = 0$ **(j)** $\prod_{i=0}^{30}(-1)^{i+1}\frac{2i+1}{2}$

(k) $\sum_{i=1}^{m}\prod_{j=1}^{n} a_i^j$ or $\sum_{i=1}^{m} a_i^{n(n+1)/2}$ **(l)** $\prod_{i=1}^{200}\sum_{j=1}^{i} j$ or $\frac{1}{2}\prod_{i=1}^{200} i(i+1)$

6.2: **(a)** 4 **(b)** 24 **(c)** -122 **(d)** -1 **(e)** 315 **(f)** -219 **(g)** 20

6.3: **(a)** 20 **(b)** 26 **(c)** 30 **(d)** 6

6.4: **(a)** $3n$ **(b)** 1 **(c)** n **(d)** $n+2$

6.5: **(a)** k **(b)** $s+1$ **(c)** 26 **(d)** 15 **(e)** x

6.6: 1

6.7: -1

6.8: $-n$

6.9: 2^n

6.10: x

6.11: -2044

6.12: **(a)** $an + \frac{dn(n-1)}{2}$ **(b)** $\frac{n(a_1+a_n)}{2}$ **(c)** 2703 **(d)** 1219 **(e)** -1273

6.13: **(a)** $\frac{g(r^n-1)}{r-1}$ **(b)** $\frac{g_n r - g_1}{r-1}$ **(c)** 1022 **(d)** 765 **(e)** 341

6.14: **(a)** $\frac{n(n^2-1)}{24}$ **(b)** 575 **(c)** 148 **(d)** $\frac{a_1+a_n}{2}$

6.15: **(a)** 44

6.16: **(a)** $n!$ **(b)** $\sqrt{2^{n+1}}$ **(c)** $3\sqrt{2^{n-1}}$ **(d)** $g\sqrt{r^{n-1}}$ **(e)** $\sqrt{g_1 g_n}$ **(g)** 2

6.17: 506

6.18: **(a)** 2520 **(b)** $n!$ **(c)** 120

6.19: **(a)** 21

The answer to 6.20 is on the next page.

6.20:

(a) $a^2 + 2ab + b^2$ and $a^3 + 3a^2b + 3ab^2 + b^3$

(b) $a^4 + 4a^3b + 6a^2b^2 + 4ab^3 + b^4$

(c) $a^5 + 5a^4b + 10a^3b^2 + 10a^2b^3 + 5ab^4 + b^5$

(e) $a^6 + 6a^5b + 15a^4b^2 + 20a^3b^3 + 15a^2b^4 + 6ab^5 + b^6$

Solutions for Chapter 1 Exercises

Exercise 1.1

(a)
$$2 \cdot 2 \cdot 3 + 3(2+3) = \boxed{27}$$

(b)
$$-(-3) - (-2)(3(-3) - (-2)) = 3 + 2(-9+2) = 3 - 14 = \boxed{-11}$$

(c)
$$2(-4-2) - 3(-4-4) = 2(-6) - 3(-8) = -12 + 24 = \boxed{12}$$

(d)
$$-(-3)(-2 - (-3) - 2(-4)) = 3(-2+3+8) = 3 \cdot 9 = \boxed{27}$$

Exercise 1.2

(a)
$$-(-1) - (-2) - (-1)^2 - (-2)^2 = 1 + 2 - 1 - 4 = \boxed{-2}$$

(b)
$$(-2)(-3)((-2)^2 - (-3)^2 - (-4)^2) = 6(4 - 9 - 16) = 6(-21) = \boxed{-126}$$

(c)
$$2^2 - (-2)^2 - 2^2(-2)^2(-3)^2 = 4 - 4 - 144 = \boxed{-144}$$

(d)
$$(-2-3)^2 - (-3)((-2)^2 + 3^2) = 25 + 3(4+9) = \boxed{64}$$

Exercise 1.3

$$-(-(-2) + 2^2 - (-1)^3)^2 = -(2 + 4 + 1)^2 - 7^2 = \boxed{-49}$$

Exercise 1.4

$$-\left(-x + \frac{x}{y}\right) = -\left(-(-4) - \frac{4}{2}\right) = -(4 - 2) = \boxed{-2}$$

Exercise 1.5

$$(x^2 - y)(y^2 - x) = ((-2)^2 - (-3))((-3)^2 - (-2)) = (4 + 3)(9 + 2) = \boxed{77}$$

Exercise 1.6

$$((-1)^2 - (-1)^{2-1}) \cdot (-1) \cdot 2 = \boxed{-4}$$

Exercise 1.7

(a)

$$1 \cdot (-1)^2 \cdot 2^3 \cdot (-2)^4 = \boxed{128}$$

(b)

$$\sqrt{-1 - 2 - 3 + 10} = \sqrt{4} = \boxed{2}$$

(c)

$$\sqrt{25\sqrt{8\sqrt{4}}} = \sqrt{25\sqrt{8 \cdot 2}} = \sqrt{25\sqrt{16}} = \sqrt{25 \cdot 4} = \sqrt{100} = \boxed{10}$$

Exercise 1.8

$$-\frac{1}{\left(-\frac{1}{2}\right)^3}\left(\left(\frac{1}{2}\right)^3 + \frac{\left(-\frac{1}{2}\right)^3}{\left(\frac{1}{2}\right)}\right) = 8\left(\frac{1}{8} - \frac{1}{8} \cdot \frac{2}{1}\right) = 8\left(-\frac{1}{8}\right) = \boxed{-1}$$

Exercise 1.9

$$\left(\frac{5 \cdot 2 + 7}{11 - 2} \cdot \frac{2 + \left(\frac{1}{2}\right)^2}{2 - \left(-\frac{1}{2}\right)^3}\right)^2 = \left(\frac{17}{9} \cdot \frac{2 + \frac{1}{4}}{2 + \frac{1}{8}}\right)^2 = \left(\frac{17}{9} \cdot \frac{9}{4} \cdot \frac{8}{17}\right)^2 = 2^2 = \boxed{4}$$

Exercise 1.10

$$\sqrt{-2(1 - (-2))(1 - (-2)^2)(1 - (-2)^3) - 11 \cdot 2 + 2^2}$$
$$= \sqrt{-2 \cdot 3 \cdot (-3) \cdot 9 - 22 + 4} = \sqrt{144} = \boxed{12}$$

Exercise 1.11

$$B = \frac{3(1)^2 - 2(-2)}{2} = \frac{3+4}{2} = \frac{7}{2}$$

$$C = \frac{4(-2)}{2} + 1 = -4 + 1 = -3$$

$$A = -2(-3)\left(\frac{2 \cdot \frac{7}{2}}{\frac{7}{2} - 2} - \frac{-3}{-3+1}\right) = 6\left(\frac{14}{3} - \frac{3}{2}\right) = 6 \cdot \frac{19}{6} = \boxed{19}$$

Exercise 1.12

$$A_1 = \frac{2 \cdot (-1)}{-2} = 1$$

$$A_2 = \frac{2 + (-1)}{-2} = -\frac{1}{2}$$

$$A = \frac{3 \cdot 1}{1 - 2} + \frac{6 \cdot 1 \cdot \left(-\frac{1}{2}\right)}{-\frac{1}{2} + 2} = -3 - \frac{6}{3} = -3 - 2 = \boxed{-5}$$

Exercise 1.13

$$A_1 = \frac{(-1)(-1)^2 + 2 \cdot 2}{1(-2)} = \frac{3}{-2} = -\frac{3}{2}$$

$$A_2 = \frac{(-2)2}{1(-2)} - 1 = 2 - 1 = 1$$

$$A = \sqrt{35\left(\frac{2\left(-\frac{3}{2}\right)^2 + 5\left(-\frac{3}{2}\right)}{-\frac{3}{2} - 2} - \frac{1}{1 - \left(-\frac{3}{2}\right)}\right)} = \sqrt{35\left(\frac{6}{7} - \frac{2}{5}\right)} = \sqrt{35 \cdot \frac{16}{35}} = \boxed{4}$$

Exercise 1.14

$$\Gamma = \frac{(-1)(-1)^3 + (-2)2}{(-2)1} = \frac{3}{2}$$

$$\beta = 3\left(\frac{1(-2)}{(-2)1}\right)^2 = 3(1)^2 = 3$$

$$A = \frac{2\left(\frac{3}{2}\right)^2\left(\frac{3}{2} - 1\right)}{\frac{3}{2} - 3}\sqrt{\frac{3+1}{3^2}} = -\frac{3}{2} \cdot \frac{2}{3} = \boxed{-1}$$

Exercise 1.15

$$\Theta = \frac{2 \cdot 1 + (-2)^2}{(-1)(-2)} = \frac{6}{2} = 3$$

$$\Phi = \frac{5(3)^3 - 2\sqrt{-(-1)}}{-2} = -\frac{133}{2} = -66.5$$

$$\Delta = \frac{(-1)}{2} - \frac{\sqrt{4}}{\sqrt{9}} = -\frac{1}{2} - \frac{2}{3} = -\frac{7}{6}$$

$$A = \frac{(3^2 + (-2)^2)\sqrt{3 - (-66.5)} - 20.5}{(-2)^2(2^6 - 1)\left(\left(-\frac{7}{6}\right)^2 - 1\right)} + \frac{3 \cdot 3}{\sqrt{9} - \sqrt{4}} = \frac{13\sqrt{49}}{91} + 9 = 1 + 9 = \boxed{10}$$

Exercise 1.16

$$\frac{4}{2} + \frac{9}{3} + \frac{8}{4} + \frac{10}{5} + \frac{6}{6} + \frac{14}{7} = 2 + 3 + 2 + 2 + 1 + 2 = \boxed{12}$$

Exercise 1.17

$$1 \cdot 2 \cdot 3 \cdot 4 \cdot 5 = \boxed{120}$$

Exercise 1.18

$$\frac{2^{6-1}(-1)(2)(2)(-1)(2)(-1)}{(-1-1)^2(2-1)^2(2-1)^2(-1-1)^2(2-1)^2(-1-1)^2} = -\frac{256}{64} = \boxed{-4}$$

Exercise 1.19

$$(-2 + (-1)^2 + (-1)^3 + (-1)^4 + (-1)^5)^5 = (-2 + 1 - 1 + 1 - 1)^5 = (-2)^5 = \boxed{-32}$$

Exercise 1.20

$$\frac{(-2)^5}{(-2)^3} + \frac{(-1)^5}{(-1)^3} + \frac{(-2)^5}{(-2)^3} + \frac{(-1)^5}{(-1)^3} + \frac{(2)^5}{(-1)^3} = 4 + 1 + 4 + 1 - 32 = \boxed{-22}$$

Exercise 1.21

The expression to represent the total cost C of the stuffed animals sold is:

$$C = ax + by + 2(a + b)$$

$$150 \cdot 15 + 230 \cdot 17 + 2 \cdot (150 + 230) = 2250 + 3910 + 2 \cdot 380 = \boxed{6920}$$

Exercise 1.22

The area of the yard is given by:

$$S = s^2$$

The area of the pool is given by:

$$T = t^2$$

The difference between the areas is:

$$D = S - T = s^2 - t^2$$

To find price, we must multiply this area by the price per square meter:

$$P = x(s^2 - t^2)$$

$$P = 30(11^2 - 9^2) = \boxed{\$1200}$$

Exercise 1.23

The area of one large pizza is:

$$A_1 = \pi r^2$$

The area of two smaller pizzas is the sum of the areas of each pizza:

$$A_2 = \pi r_1^2 + \pi r_2^2$$

The difference in area between one large pizza and two smaller pizzas is:

$$A_1 - A_2 = \pi r^2 - (\pi r_1^2 + \pi r_2^2)$$

Substituting $r = 8$, $r_1 = 6$, and $r_2 = 5$, we get:

$$A_1 - A_2 = 3.14 \cdot 8^2 - (3.14 \cdot 6^2 + 3.14 \cdot 5^2) = \boxed{9.42 \, in^2}$$

One pizza with $r = 8$ inches is more cost-effective than two pizzas with $r_1 = 6$ inches, and $r_2 = 5$ inches. Note that in later chapters you will learn techniques to simplify the algebraic expression for $A_1 - A_2$:

$$A_1 - A_2 = \pi(r^2 - r_1^2 - r_2^2)$$

Exercise 1.24

The volume of the rectangular box is given by:

$$V_{box} = xyz$$

The volume of each cube part is:

$$V_{cube\,part} = \frac{z^3}{m}$$

The total volume of the shape is:

$$V_{tank} = V_{box} + nV_{cube\,part} = xyz + \frac{nz^3}{m}$$

Plugging in the given values, we get:

$$1 \cdot 0.8 \cdot 0.6 + \frac{2(0.6)^3}{4} = \boxed{0.588\,m^3}$$

Exercise 1.25

The car will travel vt kilometers and will consume $\frac{vtf}{100}$ liters of fuel. Since the price per liter is δ, the algebraic expression for the price is:

$$P = \frac{vtf\delta}{100}$$

Substituting the values, we obtain:

$$P = \frac{80 \cdot 3 \cdot 9 \cdot 1.5}{100} = \boxed{\$32.4}$$

Solutions for Chapter 2 Exercises

Exercise 2.1

(a)
$$(3x)^2 = 3^2 x^2 = \boxed{9x^2}$$

(b)
$$(4y)^3 = 4^3 y^3 = \boxed{64y^3}$$

(c)
$$(5z)^4 = 5^4 z^4 = \boxed{625z^4}$$

(d)
$$(2\alpha)^3 = 2^3 \alpha^3 = \boxed{8\alpha^3}$$

(e)
$$(3\beta)^4 = 3^4 \beta^4 = \boxed{81\beta^4}$$

(f)
$$(4\gamma)^2 = 4^2 \gamma^2 = \boxed{16\gamma^2}$$

(g)
$$(2\delta)^4 = 2^4 \delta^4 = \boxed{16\delta^4}$$

(h)
$$(-5\epsilon)^4 = (-5)^4 \epsilon^4 = \boxed{625\epsilon^4}$$

(i)
$$(-4A)^3 = (-4)^3 A^3 = \boxed{-64A^3}$$

(j)
$$(-2\eta)^3 = (-2)^3 \eta^3 = \boxed{-8\eta^3}$$

(k)
$$(-3\theta)^4 = (-3)^4 \theta^4 = \boxed{81\theta^4}$$

(l)
$$(-4w)^2 = (-4)^2 w^2 = \boxed{16w^2}$$

Exercise 2.2

(a)
$$(2xy)^5 = 2^5 x^5 y^5 = \boxed{32x^5 y^5}$$

(b)
$$(-3yz)^4 = (-3)^4 y^4 z^4 = \boxed{81y^4 z^4}$$

(c)
$$(-5xz)^4 = (-5)^4 x^4 z^4 = \boxed{625x^4 z^4}$$

(d)
$$(5\alpha\beta)^4 = 5^4 \alpha^4 \beta^4 = \boxed{625\alpha^4 \beta^4}$$

(e)
$$(-2vw)^3 = (-2)^3 v^3 w^3 = \boxed{-8v^3 w^3}$$

(f)
$$(3mn)^5 = 3^5 m^5 n^5 = \boxed{243m^5 n^5}$$

(g)
$$(4\delta\epsilon)^2 = 4^2 \delta^2 \epsilon^2 = \boxed{16\delta^2 \epsilon^2}$$

(h)
$$(-2st)^5 = (-2)^5 s^5 t^5 = \boxed{-32s^5 t^5}$$

(i)
$$(-3DF)^4 = (-3)^4 D^4 F^4 = \boxed{81D^4 F^4}$$

(j)
$$(6BR)^3 = 6^3 B^3 R^3 = \boxed{216B^3 R^3}$$

(k)
$$(5\alpha\phi\theta)^2 = 5^2 \alpha^2 \phi^2 \theta^2 = \boxed{25\alpha^2 \phi^2 \theta^2}$$

(l)
$$(-4dsrw)^5 = (-4)^5 d^5 s^5 r^5 w^5 = \boxed{-1024d^5 s^5 r^5 w^5}$$

Exercise 2.3

(a)
$$(x^2)^3 = x^{2 \cdot 3} = \boxed{x^6}$$

(b)
$$(y^3)^4 = y^{3 \cdot 4} = \boxed{y^{12}}$$

(c)
$$(z^4)^2 = z^{4 \cdot 2} = \boxed{z^8}$$

(d)
$$(\alpha^2)^3 = \alpha^{2 \cdot 3} = \boxed{\alpha^6}$$

(e)
$$(\beta^3)^4 = \beta^{3 \cdot 4} = \boxed{\beta^{12}}$$

(f)
$$(\gamma^4)^2 = \gamma^{4 \cdot 2} = \boxed{\gamma^8}$$

(g)
$$(\delta^2)^3 = \delta^{2 \cdot 3} = \boxed{\delta^6}$$

(h)
$$(\epsilon^3)^4 = \epsilon^{3 \cdot 4} = \boxed{\epsilon^{12}}$$

(i)
$$(W^4)^2 = W^{4 \cdot 2} = \boxed{W^8}$$

(j)
$$((x^2)^3)^4 = (x^{2 \cdot 3})^4 = (x^6)^4 = x^{6 \cdot 4} = \boxed{x^{24}}$$

(k)
$$((v^3)^4)^2 = (v^{3 \cdot 4})^2 = (v^{12})^2 = v^{12 \cdot 2} = \boxed{v^{24}}$$

(l)
$$((s^4)^2)^3 = (s^{4 \cdot 2})^3 = (s^8)^3 = s^{8 \cdot 3} = \boxed{s^{24}}$$

Exercise 2.4

(a)
$$(2a^2b^3)^2 = 2^2(a^2)^2(b^3)^2 = 4a^{2\cdot2}b^{3\cdot2} = \boxed{4a^4b^6}$$

(b)
$$(3c^2d^3e^4)^2 = 3^2(c^2)^2(d^3)^2(e^4)^2 = 9c^{2\cdot2}d^{3\cdot2}e^{4\cdot2} = \boxed{9c^4d^6e^8}$$

(c)
$$(-2i^4j^3k^2)^2 = (-2)^2(i^4)^2(j^3)^2(k^2)^2 = 4i^{4\cdot2}j^{3\cdot2}k^{2\cdot2} = \boxed{4i^8j^6k^4}$$

(d)
$$(4A^2B^3C^4)^2 = 4^2(A^2)^2(B^3)^2(C^4)^2 = 16A^{2\cdot2}B^{3\cdot2}C^{4\cdot2} = \boxed{16A^4B^6C^8}$$

(e)
$$(-4o^3p^2q^4)^2 = (-4)^2(o^3)^2(p^2)^2(q^4)^2 = 16o^{3\cdot2}p^{2\cdot2}q^{4\cdot2} = \boxed{16o^6p^4q^8}$$

(f)
$$(5\alpha^2\beta^3\gamma^4)^3 = 5^3(\alpha^2)^3(\beta^3)^3(\gamma^4)^3 = 125\alpha^{2\cdot3}\beta^{3\cdot3}\gamma^{4\cdot3} = \boxed{125\alpha^6\beta^9\gamma^{12}}$$

(g)
$$(-5\delta^3\epsilon^2\zeta^4)^3 = (-5)^3(\delta^3)^3(\epsilon^2)^3(\zeta^4)^3 = -125\delta^{3\cdot3}\epsilon^{2\cdot3}\zeta^{4\cdot3} = \boxed{-125\delta^9\epsilon^6\zeta^{12}}$$

(h)
$$(-3r^2s^3t^4u^2)^3 = (-3)^3(r^2)^3(s^3)^3(t^4)^3(u^2)^3 = -27r^{2\cdot3}s^{3\cdot3}t^{4\cdot3}u^{2\cdot3} = \boxed{-27r^6s^9t^{12}u^6}$$

(i)
$$(3l^2m^3n^4)^4 = 3^4(l^2)^4(m^3)^4(n^4)^4 = 81l^{2\cdot4}m^{3\cdot4}n^{4\cdot4} = \boxed{81l^8m^{12}n^{16}}$$

(j)
$$(2a^2b^3)^4 = 2^4(a^2)^4(b^3)^4 = 16a^{2\cdot4}b^{3\cdot4} = \boxed{16a^8b^{12}}$$

(k)
$$(-4f^3g^2h^4)^4 = (-4)^4(f^3)^4(g^2)^4(h^4)^4 = 256f^{3\cdot4}g^{2\cdot4}h^{4\cdot4} = \boxed{256f^{12}g^8h^{16}}$$

(l)
$$(-3D^4E^3F^2G^4)^4 = (-3)^4(D^4)^4(E^3)^4(F^2)^4(G^4)^4$$
$$= 81D^{4\cdot4}E^{3\cdot4}F^{2\cdot4}G^{4\cdot4} = \boxed{81D^{16}E^{12}F^8G^{16}}$$

Exercise 2.5

(a)
$$(x^{-2}y^3)^2 = (x^{-2})^2(y^3)^2 = x^{-4}y^6 = \boxed{\dfrac{y^6}{x^4}}$$

(b)
$$(z^3w^{-1})^3 = (z^3)^3(w^{-1})^3 = z^9w^{-3} = \boxed{\dfrac{z^9}{w^3}}$$

(c)
$$(r^2s^{-3})^4 = (r^2)^4(s^{-3})^4 = r^8s^{-12} = \boxed{\dfrac{r^8}{s^{12}}}$$

(d)
$$(u^3v^{-2}w)^2 = (u^3)^2(v^{-2})^2(w)^2 = u^6v^{-4}w^2 = \boxed{\dfrac{u^6w^2}{v^4}}$$

(e)
$$(a^{-3}b^2)^3 = (a^{-3})^3(b^2)^3 = a^{-9}b^6 = \boxed{\dfrac{b^6}{a^9}}$$

(f)
$$(c^{-1}d^3e^{-2})^2 = (c^{-1})^2(d^3)^2(e^{-2})^2 = c^{-2}d^6e^{-4} = \boxed{\dfrac{d^6}{c^2e^4}}$$

(g)
$$(-2x^2y^{-3})^2 = (-2)^2(x^2)^2(y^{-3})^2 = 4x^4y^{-6} = \boxed{\dfrac{4x^4}{y^6}}$$

(h)
$$(3m^{-1}n^2)^3 = (3)^3(m^{-1})^3(n^2)^3 = 27m^{-3}n^6 = \boxed{\dfrac{27n^6}{m^3}}$$

(i)
$$(-4p^3q^{-2})^3 = (-4)^3(p^3)^3(q^{-2})^3 = -64p^9q^{-6} = \boxed{-\dfrac{64p^9}{q^6}}$$

(j)
$$(2\alpha^2\beta^{-3})^4 = (2)^4(\alpha^2)^4(\beta^{-3})^4 = 16\alpha^8\beta^{-12} = \boxed{\dfrac{16\alpha^8}{\beta^{12}}}$$

(k)
$$(-3S^3T^{-2}Q)^2 = (-3)^2(S^3)^2(T^{-2})^2(Q)^2 = 9S^6T^{-4}Q^2 = \boxed{\dfrac{9S^6Q^2}{T^4}}$$

(l)

$$(4\phi^{-1}\theta^2\omega^3)^{-1} = 4^{-1}(\phi^{-1})^{-1}(\theta^2)^{-1}(\omega^3)^{-1} = \frac{1}{4}\phi^1\theta^{-2}\omega^{-3} = \boxed{\frac{\phi}{4\theta^2\omega^3}}$$

Exercise 2.6

(a)

$$s^{\frac{2}{3}} = \boxed{\sqrt[3]{s^2}}$$

(b)

$$t^{0.8} = t^{\frac{4}{5}} = \boxed{\sqrt[5]{t^4}}$$

(c)

$$u^{0.75} = u^{\frac{3}{4}} = \boxed{\sqrt[4]{u^3}}$$

(d)

$$w^{\frac{5}{3}}x^{\frac{2}{7}} = \boxed{\sqrt[3]{w^5}\sqrt[7]{x^2}}$$

(e)

$$y^{0.8}z^{0.5} = y^{\frac{4}{5}}z^{\frac{1}{2}} = \boxed{\sqrt[5]{y^4}\sqrt{z}}$$

(f)

$$\alpha^{\frac{3}{5}} + \beta^{\frac{7}{10}} = \boxed{\sqrt[5]{\alpha^3} + \sqrt[10]{\beta^7}}$$

(g)

$$\gamma^{\frac{6}{7}}\delta^{\frac{2}{3}} = \boxed{\sqrt[7]{\gamma^6}\sqrt[3]{\delta^2}}$$

(h)

$$a^{0.5} + b^{0.75} = a^{\frac{1}{2}} + b^{\frac{3}{4}} = \boxed{\sqrt{a} + \sqrt[4]{b^3}}$$

(i)

$$\eta^{\frac{5}{6}}\theta^{\frac{1}{3}} = \boxed{\sqrt[6]{\eta^5}\sqrt[3]{\theta}}$$

(j)

$$l^{\frac{2}{5}}m^{\frac{3}{7}} = \boxed{\sqrt[5]{l^2}\sqrt[7]{m^3}}$$

(k)

$$\lambda^{\frac{4}{3}}\mu^{\frac{5}{8}}\nu^{\frac{2}{5}} = \boxed{\sqrt[3]{\lambda^4}\sqrt[8]{\mu^5}\sqrt[5]{\nu^2}}$$

(l)

$$G^{\frac{1}{4}} + H^{\frac{2}{3}} + J^{\frac{7}{10}} = \boxed{\sqrt[4]{G} + \sqrt[3]{H^2} + \sqrt[10]{J^7}}$$

Exercise 2.7

(a)
$$\sqrt[5]{A} = \boxed{A^{\frac{1}{5}}}$$

(b)
$$\sqrt[6]{y^5} = \boxed{y^{\frac{5}{6}}}$$

(c)
$$\sqrt[7]{z^2} = \boxed{z^{\frac{2}{7}}}$$

(d)
$$\sqrt[3]{q^2} = \boxed{q^{\frac{2}{3}}}$$

(e)
$$\sqrt[4]{r^5} = \boxed{r^{\frac{5}{4}}}$$

(f)
$$\sqrt[9]{u^2} = \boxed{u^{\frac{2}{9}}}$$

(g)
$$\sqrt[3]{v}\sqrt[4]{w} = \boxed{v^{\frac{1}{3}} w^{\frac{1}{4}}}$$

(h)
$$\sqrt[5]{x^3} + \sqrt[3]{y^2} = \boxed{x^{\frac{3}{5}} + y^{\frac{2}{3}}}$$

(i)
$$\sqrt[6]{w^7}\sqrt[2]{z} = \boxed{w^{\frac{7}{6}} z^{\frac{1}{2}}}$$

(j)
$$\sqrt[7]{\alpha^2}\sqrt[7]{\beta^3}\sqrt[7]{\gamma^4} = \boxed{\alpha^{\frac{2}{7}} \beta^{\frac{3}{7}} \gamma^{\frac{4}{7}}}$$

(k)
$$\sqrt[3]{b^5} + \sqrt[5]{d^7} + \sqrt[7]{e^9} = \boxed{b^{\frac{5}{3}} + d^{\frac{7}{5}} + e^{\frac{9}{7}}}$$

(l)
$$\sqrt[4]{c} + \sqrt[3]{d^2}\sqrt[2]{e^3} = \boxed{c^{\frac{1}{4}} + d^{\frac{2}{3}} e^{\frac{3}{2}}}$$

Exercise 2.8

(a)
$$\left(a^{\frac{1}{2}}b^{\frac{1}{3}}\right)^2 = a^{\frac{1}{2}\cdot 2}b^{\frac{1}{3}\cdot 2} = ab^{\frac{2}{3}} = \boxed{a\sqrt[3]{b^2}}$$

(b)
$$\left(\Delta^{\frac{1}{3}}\Gamma^{\frac{1}{2}}\right)^3 = \Delta^{\frac{1}{3}\cdot 3}\Gamma^{\frac{1}{2}\cdot 3} = \Delta\Gamma^{\frac{3}{2}} = \boxed{\Delta\sqrt{\Gamma^3}}$$

(c)
$$\left(e^{\frac{1}{2}}f^{\frac{1}{3}}\right)^4 = e^{\frac{1}{2}\cdot 4}f^{\frac{1}{3}\cdot 4} = e^2 f^{\frac{4}{3}} = \boxed{e^2\sqrt[3]{f^4}}$$

(d)
$$\left(g^{\frac{1}{3}}h^{\frac{1}{2}}i^{\frac{1}{5}}\right)^2 = g^{\frac{1}{3}\cdot 2}h^{\frac{1}{2}\cdot 2}i^{\frac{1}{5}\cdot 2} = g^{\frac{2}{3}}hi^{\frac{2}{5}} = \boxed{\sqrt[3]{g^2}h\sqrt[5]{i^2}}$$

(e)
$$\left(-3j^{\frac{1}{3}}k^{\frac{1}{2}}\right)^3 = (-3)^3 j^{\frac{1}{3}\cdot 3}k^{\frac{1}{2}\cdot 3} = -27jk^{\frac{3}{2}} = \boxed{-27j\sqrt{k^3}}$$

(f)
$$\left(2l^{\frac{1}{2}}m^{\frac{1}{3}}n^{\frac{1}{4}}\right)^2 = 2^2 l^{\frac{1}{2}\cdot 2}m^{\frac{1}{3}\cdot 2}n^{\frac{1}{4}\cdot 2} = 4lm^{\frac{2}{3}}n^{\frac{1}{2}} = \boxed{4l\sqrt[3]{m^2}\sqrt{n}}$$

(g)
$$\left(-\alpha^{-\frac{1}{2}}\beta^{\frac{1}{2}}\right)^{\frac{2}{3}} = (-1)^{\frac{2}{3}}\alpha^{-\frac{1}{2}\cdot\frac{2}{3}}\beta^{\frac{1}{2}\cdot\frac{2}{3}} = \left((-1)^2\right)^{\frac{1}{3}}\alpha^{-\frac{2}{6}}\beta^{\frac{2}{6}} = \alpha^{-\frac{1}{3}}\beta^{\frac{1}{3}} = \frac{\beta^{\frac{1}{3}}}{\alpha^{\frac{1}{3}}} = \boxed{\sqrt[3]{\frac{\beta}{\alpha}}}$$

(h)
$$\left(q^{\frac{1}{2}}r^{-\frac{1}{3}}\right)^{\frac{5}{3}} = q^{\frac{1}{2}\cdot\frac{5}{3}}r^{-\frac{1}{3}\cdot\frac{5}{3}} = q^{\frac{5}{6}}r^{-\frac{5}{9}} = \frac{q^{\frac{5}{6}}}{r^{\frac{5}{9}}} = \boxed{\frac{\sqrt[6]{q^5}}{\sqrt[9]{r^5}}}$$

(i)
$$\left(-W^{\frac{1}{3}}Z^{-\frac{1}{2}}\right)^{\frac{7}{3}} = (-1)^{\frac{7}{3}}W^{\frac{1}{3}\cdot\frac{7}{3}}Z^{-\frac{1}{2}\cdot\frac{7}{3}} = -W^{\frac{7}{9}}Z^{-\frac{7}{6}} = -\frac{W^{\frac{7}{9}}}{Z^{\frac{7}{6}}} = \boxed{-\frac{\sqrt[9]{W^7}}{\sqrt[6]{Z^7}}}$$

(j)
$$\left(u^{-\frac{1}{2}}v^{\frac{1}{3}}\right)^{\frac{5}{2}} = u^{-\frac{1}{2}\cdot\frac{5}{2}}v^{\frac{1}{3}\cdot\frac{5}{2}} = u^{-\frac{5}{4}}v^{\frac{5}{6}} = \frac{v^{\frac{5}{6}}}{u^{\frac{5}{4}}} = \boxed{\frac{\sqrt[6]{v^5}}{\sqrt[4]{u^5}}}$$

(k)
$$\left(-w^{-\frac{1}{3}}x^{\frac{1}{2}}y^{\frac{1}{4}}\right)^{-\frac{4}{3}} = (-1)^{-\frac{4}{3}}w^{-\frac{1}{3}\cdot-\frac{4}{3}}x^{\frac{1}{2}\cdot-\frac{4}{3}}y^{\frac{1}{4}\cdot-\frac{4}{3}} = w^{\frac{4}{9}}x^{-\frac{2}{3}}y^{-\frac{1}{3}} = \frac{w^{\frac{4}{9}}}{x^{\frac{2}{3}}y^{\frac{1}{3}}} = \frac{w^{\frac{4}{9}}}{(x^2 y)^{\frac{1}{3}}} = \boxed{\frac{\sqrt[9]{w^4}}{\sqrt[3]{x^2 y}}}$$

(l)

$$\left(\gamma^{-\frac{1}{2}} A^{-\frac{1}{3}} B^{\frac{1}{4}}\right)^{-\frac{5}{4}} = \gamma^{-\frac{1}{2} \cdot -\frac{5}{4}} A^{-\frac{1}{3} \cdot -\frac{5}{4}} B^{\frac{1}{4} \cdot -\frac{5}{4}} = \gamma^{\frac{5}{8}} A^{\frac{5}{12}} B^{-\frac{5}{16}} = \frac{\gamma^{\frac{5}{8}} A^{\frac{5}{12}}}{B^{\frac{5}{16}}} = \boxed{\frac{\sqrt[8]{\gamma^5} \sqrt[12]{A^5}}{\sqrt[16]{B^5}}}$$

Exercise 2.9

(a)
$$a^3 a^2 = a^{3+2} = \boxed{a^5}$$

(b)
$$c^{-5} c^3 = c^{-5+3} = c^{-2} = \boxed{\frac{1}{c^2}}$$

(c)
$$\alpha^3 \alpha^{-2} = \alpha^{3-2} = \boxed{\alpha}$$

(d)
$$d^4 d^{-6} = d^{4-6} = d^{-2} = \boxed{\frac{1}{d^2}}$$

(e)
$$E^2 E^{-4} = E^{2-4} = E^{-2} = \boxed{\frac{1}{E^2}}$$

(f)
$$e^4 e^{-1} = e^{4-1} = e^3 = \boxed{e^3}$$

(g)
$$l l^{-1} g^{-2} g^5 = l^{1-1} g^{-2+5} = l^0 g^3 = \boxed{g^3}$$

(h)
$$K^{-1} H H^{-3} K = H^{1-3} K^{1-1} = H^{-2} K^0 = \boxed{\frac{1}{H^2}}$$

(i)
$$s^3 k^3 s^{-3} k^4 = k^{3+4} s^{3-3} = \boxed{k^7}$$

(j)
$$j^{-1} j^{-4} j^{-2} = j^{-1-4-2} = j^{-7} = \boxed{\frac{1}{j^7}}$$

(k)
$$\Delta^3 \Delta^{-5} \Delta^3 = \Delta^{3-5+3} = \Delta^1 = \boxed{\Delta}$$

(l)
$$\lambda^4 \lambda^2 \lambda^{-5} = \lambda^{4+2-5} = \lambda^1 = \boxed{\lambda}$$

Exercise 2.10

(a)
$$\Delta^{1.5}\Delta^{0.5} = \Delta^{\frac{3}{2}}\Delta^{\frac{1}{2}} = \Delta^{\frac{3}{2}+\frac{1}{2}} = \Delta^{\frac{4}{2}} = \boxed{\Delta^2}$$

(b)
$$g^{-\frac{5}{3}}g^{\frac{2}{3}} = g^{-\frac{5}{3}+\frac{2}{3}} = g^{-\frac{3}{3}} = g^{-1} = \boxed{\frac{1}{g}}$$

(c)
$$R^{\frac{3}{4}}R^{-\frac{1}{4}} = R^{\frac{3}{4}-\frac{1}{4}} = R^{\frac{2}{4}} = R^{\frac{1}{2}} = \boxed{\sqrt{R}}$$

(d)
$$z^{\frac{4}{5}}z^{-\frac{6}{5}} = z^{\frac{4}{5}-\frac{6}{5}} = z^{-\frac{2}{5}} = \boxed{\sqrt[5]{\frac{1}{z^2}}}$$

(e)
$$p^{\frac{2}{3}}p^{-\frac{4}{3}} = p^{\frac{2}{3}-\frac{4}{3}} = p^{-\frac{2}{3}} = \boxed{\sqrt[3]{\frac{1}{p^2}}}$$

(f)
$$t^{\frac{4}{7}}t^{-\frac{1}{7}} = t^{\frac{4}{7}-\frac{1}{7}} = t^{\frac{3}{7}} = \boxed{\sqrt[7]{t^3}}$$

(g)
$$u^{-\frac{2}{9}}u^{\frac{5}{9}} = u^{-\frac{2}{9}+\frac{5}{9}} = u^{\frac{3}{9}} = u^{\frac{1}{3}} = \boxed{\sqrt[3]{u}}$$

(h)
$$W^{\frac{1}{2}}W^{-\frac{3}{2}} = W^{\frac{1}{2}-\frac{3}{2}} = W^{-\frac{2}{2}} = W^{-1} = \boxed{\frac{1}{W}}$$

(i)
$$Y^{\frac{3}{4}}Y^{\frac{1}{4}} = Y^{\frac{3}{4}+\frac{1}{4}} = Y^{\frac{4}{4}} = Y^1 = \boxed{Y}$$

(j)
$$Q^{-\frac{1}{5}}Q^{-\frac{4}{5}}Q^{-\frac{2}{5}} = Q^{-\frac{1}{5}-\frac{4}{5}-\frac{2}{5}} = Q^{-\frac{7}{5}} = \boxed{\sqrt[5]{\frac{1}{Q^7}}}$$

(k)
$$L^{1.5}L^{-2.5}L^{1.5} = L^{\frac{3}{2}}L^{-\frac{5}{2}}L^{\frac{3}{2}} = L^{\frac{3}{2}-\frac{5}{2}+\frac{3}{2}} = L^{\frac{1}{2}} = \boxed{\sqrt{L}}$$

(l)
$$V^{\frac{4}{9}}V^{\frac{2}{9}}V^{-\frac{5}{9}} = V^{\frac{4}{9}+\frac{2}{9}-\frac{5}{9}} = V^{\frac{1}{9}} = \boxed{\sqrt[9]{V}}$$

Exercise 2.11

(a)
$$A^{\frac{5}{6}}A^{\frac{1}{3}} = A^{\frac{5}{6}+\frac{1}{3}} = A^{\frac{5}{6}+\frac{2}{6}} = A^{\frac{7}{6}} = \boxed{\sqrt[6]{A^7}}$$

(b)
$$\beta^{-\frac{7}{4}}\beta^{\frac{2}{3}} = \beta^{-\frac{7}{4}+\frac{2}{3}} = \beta^{-\frac{21}{12}+\frac{8}{12}} = \beta^{-\frac{13}{12}} = \boxed{\sqrt[12]{\frac{1}{\beta^{13}}}}$$

(c)
$$W^{\frac{5}{8}}W^{-\frac{2}{5}} = W^{\frac{5}{8}-\frac{2}{5}} = W^{\frac{25}{40}-\frac{16}{40}} = W^{\frac{9}{40}} = \boxed{\sqrt[40]{W^9}}$$

(d)
$$\delta^{\frac{3}{10}}\delta^{-\frac{7}{5}} = \delta^{\frac{3}{10}-\frac{14}{10}} = \delta^{-\frac{11}{10}} = \boxed{\sqrt[10]{\frac{1}{\delta^{11}}}}$$

(e)
$$m^{\frac{3}{4}}m^{-\frac{7}{2}} = m^{\frac{3}{4}-\frac{14}{4}} = m^{-\frac{11}{4}} = \boxed{\sqrt[4]{\frac{1}{m^{11}}}}$$

(f)
$$\zeta^{\frac{4}{7}}\zeta^{-\frac{1}{3}} = \zeta^{\frac{4}{7}-\frac{1}{3}} = \zeta^{\frac{12}{21}-\frac{7}{21}} = \zeta^{\frac{5}{21}} = \boxed{\sqrt[21]{\zeta^5}}$$

(g)
$$G^{-\frac{2}{9}}G^{\frac{5}{6}} = G^{-\frac{2}{9}+\frac{5}{6}} = G^{-\frac{12}{54}+\frac{45}{54}} = G^{\frac{11}{18}} = \boxed{\sqrt[18]{G^{11}}}$$

(h)
$$s^{\frac{1}{2}}s^{-\frac{3}{8}} = s^{\frac{4}{8}-\frac{3}{8}} = s^{\frac{1}{8}} = \boxed{\sqrt[8]{s}}$$

(i)
$$I^{\frac{3}{4}}I^{\frac{3}{5}} = I^{\frac{15}{20}+\frac{12}{20}} = I^{\frac{27}{20}} = \boxed{\sqrt[20]{I^{27}}}$$

(j)
$$\kappa^{-\frac{1}{5}}\kappa^{-\frac{4}{7}}\kappa^{-\frac{2}{9}} = \kappa^{-\frac{1}{5}-\frac{4}{7}-\frac{2}{9}} = \kappa^{-\frac{313}{315}} = \boxed{\sqrt[315]{\frac{1}{\kappa^{313}}}}$$

(k)
$$R^{\frac{3}{2}}R^{-\frac{5}{3}}R^{\frac{3}{4}} = R^{\frac{3}{2}-\frac{5}{3}+\frac{3}{4}} = R^{\frac{18}{12}-\frac{20}{12}+\frac{9}{12}} = R^{\frac{7}{12}} = \boxed{\sqrt[12]{R^7}}$$

(l)
$$\lambda^{\frac{4}{9}}\lambda^{\frac{1}{7}}\lambda^{-\frac{5}{18}} = \lambda^{\frac{4}{9}+\frac{1}{7}-\frac{5}{18}} = \lambda^{\frac{56}{126}+\frac{18}{126}-\frac{35}{126}} = \lambda^{\frac{39}{126}} = \lambda^{\frac{13}{42}} = \boxed{\sqrt[42]{\lambda^{13}}}$$

Exercise 2.12

(a)
$$\left(u^{\frac{1}{2}}\sqrt[3]{u}\right)^2 = \left(u^{\frac{1}{2}}\right)^2\left(\sqrt[3]{u}\right)^2 = u^{\frac{2}{2}}u^{\frac{2}{3}} = u^{\frac{3}{3}+\frac{2}{3}} = u^{\frac{5}{3}} = \boxed{\sqrt[3]{u^5}}$$

(b)
$$\left(\sqrt{v}\sqrt[4]{v^3}\right)^3 = \left(\sqrt{v}\right)^3\left(\sqrt[4]{v^3}\right)^3 = v^{\frac{3}{2}}v^{\frac{9}{4}} = v^{\frac{6}{4}+\frac{9}{4}} = v^{\frac{15}{4}} = \boxed{\sqrt[4]{v^{15}}}$$

(c)
$$\left(w^{\frac{3}{2}}\sqrt[5]{w^2}\right)^{\frac{2}{3}} = \left(w^{\frac{3}{2}}\right)^{\frac{2}{3}}\left(\sqrt[5]{w^2}\right)^{\frac{2}{3}} = w^{\frac{6}{6}}w^{\frac{4}{15}} = w^{\frac{6}{6}+\frac{4}{15}} = w^{\frac{19}{15}} = \boxed{\sqrt[15]{w^{19}}}$$

(d)
$$\left(\sqrt[3]{x^2}\sqrt{x}\right)^{\frac{4}{3}} = \left(\sqrt[3]{x^2}\right)^{\frac{4}{3}}\left(\sqrt{x}\right)^{\frac{4}{3}} = x^{\frac{8}{9}}x^{\frac{4}{6}} = x^{\frac{8}{9}+\frac{4}{6}} = x^{\frac{14}{9}} = \boxed{\sqrt[9]{x^{14}}}$$

(e)
$$\left(\alpha^{\frac{3}{2}}\sqrt[4]{\alpha^3}\right)^{\frac{5}{6}} = \left(\alpha^{\frac{3}{2}}\right)^{\frac{5}{6}}\left(\sqrt[4]{\alpha^3}\right)^{\frac{5}{6}} = \alpha^{\frac{15}{12}}\alpha^{\frac{15}{24}} = \alpha^{\frac{5}{4}+\frac{5}{8}} = \alpha^{\frac{15}{8}} = \boxed{\sqrt[8]{\alpha^{15}}}$$

(f)
$$\left(\beta^{\frac{1}{4}}\sqrt[5]{\beta^2}\right)^{\frac{8}{5}} = \left(\beta^{\frac{1}{4}}\right)^{\frac{8}{5}}\left(\sqrt[5]{\beta^2}\right)^{\frac{8}{5}} = \beta^{\frac{8}{20}}\beta^{\frac{16}{25}} = \beta^{\frac{2}{5}+\frac{16}{25}} = \beta^{\frac{26}{25}} = \boxed{\sqrt[25]{\beta^{26}}}$$

(g)
$$\left(\sqrt[4]{\gamma^3}\gamma^{\frac{3}{2}}\right)^{\frac{7}{4}} = \left(\sqrt[4]{\gamma^3}\right)^{\frac{7}{4}}\left(\gamma^{\frac{3}{2}}\right)^{\frac{7}{4}} = \gamma^{\frac{21}{16}}\gamma^{\frac{21}{8}} = \gamma^{\frac{21}{16}+\frac{21}{8}} = \gamma^{\frac{63}{16}} = \boxed{\sqrt[16]{\gamma^{63}}}$$

(h)
$$\left(\sqrt{q}\sqrt[3]{q^5}\right)^{\frac{9}{10}} = (\sqrt{q})^{\frac{9}{10}}\left(\sqrt[3]{q^5}\right)^{\frac{9}{10}} = q^{\frac{9}{20}}q^{\frac{45}{30}} = q^{\frac{9}{20}+\frac{3}{2}} = q^{\frac{39}{20}} = \boxed{\sqrt[20]{q^{39}}}$$

(i)
$$\left(\sqrt[5]{\omega^4}\omega^{\frac{2}{3}}\right)^{\frac{15}{8}} = \left(\sqrt[5]{\omega^4}\right)^{\frac{15}{8}}\left(\omega^{\frac{2}{3}}\right)^{\frac{15}{8}} = \omega^{\frac{60}{40}}\omega^{\frac{30}{24}} = \omega^{\frac{3}{2}+\frac{5}{4}} = \omega^{\frac{11}{4}} = \boxed{\sqrt[4]{\omega^{11}}}$$

(j)
$$\left(\sqrt[3]{\kappa^{-5}}\sqrt[4]{\kappa^3}\sqrt[5]{\kappa^2}\right)^{-\frac{12}{5}} = \left(\sqrt[3]{\kappa^{-5}}\right)^{-\frac{12}{5}}\left(\sqrt[4]{\kappa^3}\right)^{-\frac{12}{5}}\left(\sqrt[5]{\kappa^2}\right)^{-\frac{12}{5}}$$
$$= \kappa^4\kappa^{-\frac{9}{5}}\kappa^{-\frac{24}{25}} = \kappa^{\frac{4}{1}-\frac{9}{5}-\frac{24}{25}} = \kappa^{\frac{31}{25}} = \boxed{\sqrt[25]{\kappa^{31}}}$$

(k)
$$\left(\sqrt[5]{\mu^{-4}}\mu^{-\frac{1}{3}}\sqrt[5]{\mu^{-3}}\right)^{-\frac{15}{4}} = \left(\sqrt[5]{\mu^{-4}}\right)^{-\frac{15}{4}}\left(\mu^{-\frac{1}{3}}\right)^{-\frac{15}{4}}\left(\sqrt[5]{\mu^{-3}}\right)^{-\frac{15}{4}}$$
$$= \mu^3\mu^{\frac{5}{4}}\mu^{\frac{9}{4}} = \mu^{\frac{3}{1}+\frac{5}{4}+\frac{9}{4}} = \mu^{\frac{13}{2}} = \boxed{\sqrt{\mu^{13}}}$$

(1)

$$\left(\left(\sqrt[3]{\nu^2}\sqrt[4]{\nu^5}\sqrt[3]{\nu^{-2}}\right)^{\frac{9}{4}}\right)^{\frac{4}{15}} = \left(\left(\left(\sqrt[3]{\nu^2}\right)^{\frac{9}{4}}\left(\sqrt[4]{\nu^5}\right)^{\frac{9}{4}}\left(\sqrt[3]{\nu^{-2}}\right)^{\frac{9}{4}}\right)\right)^{\frac{4}{15}}$$

$$= \left(\nu^{\frac{3}{2}}\nu^{\frac{45}{16}}\nu^{-\frac{3}{2}}\right)^{\frac{4}{15}} = \left(\nu^{\frac{3}{2}+\frac{45}{16}-\frac{3}{2}}\right)^{\frac{4}{15}} = \left(\nu^{\frac{45}{16}}\right)^{\frac{4}{15}} = \nu^{\frac{45}{16}\cdot\frac{4}{15}} = \nu^{\frac{3}{4}} = \boxed{\sqrt[4]{\nu^3}}$$

Exercise 2.13

(a)

$$\frac{(3aK^2)^3}{(6a^2K)^2} = \frac{3^3a^3(K^2)^3}{6^2(a^2)^2K^2} = \frac{27a^3K^6}{36a^4K^2} = \frac{3}{4}a^{3-4}K^{6-2} = \frac{3}{4}a^{-1}K^4 = \boxed{\frac{3K^4}{4a}}$$

(b)

$$\frac{(4r^3G)^2}{(8r^2G^2)^2} = \frac{4^2(r^3)^2G^2}{8^2(r^2)^2(G^2)^2} = \frac{16r^6G^2}{64r^4G^4} = \frac{1}{4}r^{6-4}G^{2-4} = \frac{1}{4}r^2G^{-2} = \boxed{\frac{r^2}{4G^2}}$$

(c)

$$\frac{(5T^4s^2)^3}{(10T^2s^4)^2} = \frac{5^3(T^4)^3(s^2)^3}{10^2(T^2)^2(s^4)^2} = \frac{125T^{12}s^6}{100T^4s^8} = \frac{5}{4}T^{12-4}s^{6-8} = \frac{5}{4}T^8s^{-2} = \boxed{\frac{5T^8}{4s^2}}$$

(d)

$$\left(\frac{xz^5y^3}{2z^3y^5x}\right)^3 = \left(\frac{1}{2}x^{1-1}z^{5-3}y^{3-5}\right)^3 = \left(\frac{1}{2}x^0z^2y^{-2}\right)^3 = \frac{1}{2^3}z^{2\cdot3}y^{-2\cdot3} = \boxed{\frac{z^6}{8y^6}}$$

(e)

$$108\left(\frac{J^2K^2B^6}{K^26J^6B^2}\right)^3 = 108\left(\frac{1}{6}K^{2-2}J^{2-6}B^{6-2}\right)^3 = 108\left(\frac{1}{6}K^0J^{-4}B^4\right)^3$$

$$= \frac{108}{6^3}J^{-4\cdot3}B^{4\cdot3} = \boxed{\frac{B^{12}}{2J^{12}}}$$

(f)

$$16\left(\frac{4H^3L^7}{H^7L^3}\right)^{-2} = 16\left(4H^{3-7}L^{7-3}\right)^{-2} = 16\left(4H^{-4}L^4\right)^{-2}$$

$$= 16\cdot4^{-2}H^{-4\cdot(-2)}L^{4\cdot(-2)} = \frac{16}{4^2}\cdot\frac{H^8}{L^8} = \boxed{\frac{H^8}{L^8}}$$

(g)

$$\frac{(3xy^{-2})^{-3}}{(3x^2y)^{-2}} = \frac{3^{-3}x^{-3}(y^{-2})^{-3}}{3^{-2}(x^2)^{-2}y^{-2}} = \frac{9}{27} \cdot \frac{x^{-3}y^6}{x^{-4}y^{-2}} = \frac{1}{3}x^{(-3)-(-4)}y^{6-(-2)} = \boxed{\frac{xy^8}{3}}$$

(h)

$$36\frac{(-3r^{-3}G)^{-2}}{(2r^{-2}G^2)^2} = 36\frac{(-3)^{-2}(r^{-3})^{-2}G^{-2}}{(2)^2(r^{-2})^2(G^2)^2} = 36\frac{r^6G^{-2}}{36r^{-4}G^4} = r^{6-(-4)}G^{-2-4} = \boxed{\frac{r^{10}}{G^6}}$$

(i)

$$\left(\frac{(T^4s^{-2})^3}{(2T^2s^4)^{-2}}\right)^3 = \left(\frac{T^{4\cdot3}s^{-2\cdot3}}{2^{-2}T^{2\cdot(-2)}s^{4\cdot(-2)}}\right)^3 = \left(\frac{4T^{12}s^{-6}}{T^{-4}s^{-8}}\right)^3 = \left(4T^{12-(-4)}s^{-6-(-8)}\right)^3$$

$$= \left(4T^{16}s^2\right)^3 = 4^3T^{16\cdot3}s^{2\cdot3} = \boxed{64T^{48}s^6}$$

(j)

$$\frac{(A+B)^{-7}B^{-9}}{B^{-10}(A+B)^{-7}} = (A+B)^{-7-(-7)}B^{-9-(-10)} = (A+B)^0B = \boxed{B}$$

(k)

$$\frac{(q+2p)^{-3}(q-3p)^4}{(q-3p)^3(q+2p)^{-3}} = (q+2p)^{-3-(-3)}(q-3p)^{4-3} = (q+2p)^0(q-3p)^1 = \boxed{q-3p}$$

(l)

$$\left(\frac{(r-w)^3q^4}{(r-w)^2q^{-5}(r-w)}\right)^2 = ((r-w)^{3-2-1}q^{4-(-5)})^2 = ((r-w)^0q^9)^2 = \boxed{q^{18}}$$

Exercise 2.14

(a)

$$\frac{\sqrt[3]{a^2}}{a^{\frac{5}{6}}} = a^{\frac{2}{3}}a^{-\frac{5}{6}} = a^{\frac{2}{3}-\frac{5}{6}} = a^{-\frac{1}{6}} = \frac{1}{\sqrt[6]{a}} = \boxed{\sqrt[6]{\frac{1}{a}}}$$

(b)

$$\frac{\sqrt[5]{L^3}L^{0.5}}{L^{0.75}} = \frac{\sqrt[5]{L^3}L^{\frac{1}{2}}}{L^{\frac{3}{4}}} = L^{\frac{3}{5}}L^{\frac{1}{2}}L^{-\frac{3}{4}} = L^{\frac{3}{5}+\frac{1}{2}-\frac{3}{4}} = L^{\frac{7}{20}} = \boxed{\sqrt[20]{L^7}}$$

(c)

$$\frac{\sqrt[5]{M^4}M^{\frac{2}{3}}}{\sqrt[3]{M^7}} = M^{\frac{4}{5}}M^{\frac{2}{3}}M^{-\frac{7}{3}} = M^{\frac{4}{5}+\frac{2}{3}-\frac{7}{3}} = M^{-\frac{13}{15}} = \frac{1}{\sqrt[15]{M^{13}}} = \boxed{\sqrt[15]{\frac{1}{M^{13}}}}$$

(d)

$$\frac{\sqrt[6]{\rho^5}\,\rho^{\frac{1}{3}}}{\rho^{0.3}\sqrt[3]{\rho^4}} = \frac{\sqrt[6]{\rho^5}\,\rho^{\frac{1}{3}}}{\rho^{\frac{3}{10}}\sqrt[3]{\rho^4}} = \rho^{\frac{5}{6}}\rho^{\frac{1}{3}}\rho^{-\frac{3}{10}}\rho^{-\frac{4}{3}} = \rho^{\frac{5}{6}+\frac{1}{3}-\frac{3}{10}-\frac{4}{3}} = \rho^{-\frac{7}{15}} = \frac{1}{\sqrt[15]{\rho^7}} = \boxed{\sqrt[15]{\frac{1}{\rho^7}}}$$

(e) To simplify the process, denote $A = K + 2L$:

$$\frac{\sqrt[3]{A^2}\sqrt[4]{A^3}}{A^{\frac{3}{5}}A^{\frac{5}{3}}} = A^{\frac{2}{3}}A^{\frac{3}{4}}A^{-\frac{3}{5}}A^{-\frac{5}{3}} = A^{\frac{2}{3}+\frac{3}{4}-\frac{3}{5}-\frac{5}{3}} = A^{-\frac{17}{20}} = \frac{1}{\sqrt[20]{A^{17}}} = \boxed{\sqrt[20]{\frac{1}{(K+2L)^{17}}}}$$

(f) To simplify the process, denote $A = 3\xi - \phi$:

$$\frac{\sqrt[4]{A^3}A^{\frac{1}{2}}\sqrt[6]{A^5}}{A^{\frac{7}{9}}} = A^{\frac{3}{4}}A^{\frac{1}{2}}A^{\frac{5}{6}}A^{-\frac{7}{9}} = A^{\frac{3}{4}+\frac{1}{2}+\frac{5}{6}-\frac{7}{9}} = A^{\frac{47}{36}} = \frac{1}{\sqrt[36]{A^{47}}} = \boxed{\sqrt[36]{\frac{1}{(3\xi-\phi)^{47}}}}$$

(g)

$$\left(\frac{\sqrt[5]{Z^4}Z^{\frac{2}{3}}\sqrt[4]{Z^3}}{\sqrt[5]{Z^3}}\right)^{\frac{10}{97}} = \left(Z^{\frac{4}{5}}Z^{\frac{2}{3}}Z^{\frac{3}{4}}Z^{-\frac{3}{5}}\right)^{\frac{10}{97}} = \left(Z^{\frac{4}{5}+\frac{2}{3}+\frac{3}{4}-\frac{3}{5}}\right)^{\frac{10}{97}} = Z^{\frac{97}{60}\cdot\frac{10}{97}} = \boxed{\sqrt[6]{Z}}$$

(h)

$$\sqrt[23]{\left(\frac{\pi^{\frac{35}{18}}}{\sqrt[6]{\pi^5}\pi^{\frac{1}{3}}\sqrt[3]{\pi^2}\sqrt[4]{\pi^3}}\right)^{12}} = \left(\pi^{\frac{35}{18}}\pi^{-\frac{5}{6}}\pi^{-\frac{1}{3}}\pi^{-\frac{2}{3}}\pi^{-\frac{3}{4}}\right)^{\frac{12}{23}}$$

$$= \left(\pi^{\frac{35}{18}-\frac{5}{6}-\frac{1}{3}-\frac{2}{3}-\frac{3}{4}}\right)^{\frac{12}{23}} = \pi^{-\frac{23}{36}\cdot\frac{12}{23}} = \frac{1}{\sqrt[3]{\pi}} = \boxed{\sqrt[3]{\frac{1}{\pi}}}$$

(i)

$$\left(\frac{\sqrt[3]{v^{0.4}}v^{-0.8}\sqrt[3]{v^{-2.5}}}{\sqrt[3]{v^{-0.75}}}\right)^{-2} = \left(\frac{\sqrt[3]{v^{\frac{2}{5}}}v^{-\frac{4}{5}}\sqrt[3]{v^{-\frac{5}{2}}}}{\sqrt[3]{v^{-\frac{3}{4}}}}\right)^{-2} = \left(v^{\frac{2}{15}-\frac{4}{5}-\frac{5}{6}+\frac{3}{12}}\right)^{-2}$$

$$= \left(v^{-\frac{5}{4}}\right)^{-2} = v^{\frac{5}{2}} = \boxed{\sqrt{v^5}}$$

(j)

$$\left(\frac{t^{-\frac{5}{6}}t^{\frac{2}{3}}}{\sqrt[4]{t^5}t^{-\frac{2}{3}}\sqrt[3]{t^2}\sqrt[4]{t^{-3}}t^{-\frac{5}{3}}}\right)^4 = \left(\frac{t^{-\frac{5}{6}}t^{\frac{2}{3}}}{t^{\frac{5}{4}}t^{-\frac{2}{3}}t^{\frac{2}{3}}t^{-\frac{3}{4}}t^{-\frac{5}{12}}}\right)^4 = \left(t^{-\frac{5}{6}+\frac{2}{3}-\frac{5}{4}+\frac{2}{3}-\frac{2}{3}+\frac{3}{4}+\frac{5}{12}}\right)^4 = \left(t^{-\frac{1}{4}}\right)^4 = \boxed{\frac{1}{t}}$$

Exercise 2.15

1.

$$\sqrt[6]{p^5}\;\sqrt[35]{\frac{\sqrt[6]{3}\left(\left((3p)^{-\frac{1}{2}}\right)^3\right)^{-\frac{1}{3}}}{p^{-6}\sqrt[3]{(3p)^2}}} = p^{\frac{5}{6}}\;\sqrt[35]{\frac{3^{\frac{1}{6}}(3p)^{\frac{3}{6}}}{p^{-6}(3p)^{\frac{2}{3}}}} = p^{\frac{5}{6}}\;\sqrt[35]{3^{\frac{1}{6}}p^6(3p)^{\frac{1}{2}-\frac{2}{3}}} = p^{\frac{5}{6}}\;\sqrt[35]{3^{\frac{1}{6}}p^6(3p)^{-\frac{1}{6}}}$$

$$= p^{\frac{5}{6}}\;\sqrt[35]{3^{\frac{1}{6}}p^63^{-\frac{1}{6}}p^{-\frac{1}{6}}} = p^{\frac{5}{6}}\;\sqrt[35]{3^0p^{6-\frac{1}{6}}} = p^{\frac{5}{6}}\;\sqrt[35]{p^{\frac{35}{6}}} = p^{\frac{5}{6}}\left(p^{\frac{35}{6}}\right)^{\frac{1}{35}} = p^{\frac{5}{6}}p^{\frac{1}{6}} = p^{\frac{5}{6}+\frac{1}{6}} = p$$

2.

$$\left(\frac{(2q)^{\frac{3}{2}}}{4q^4}\right)^{-2} = \left(\frac{4q^4}{2^{\frac{3}{2}}q^{\frac{3}{2}}}\right)^2 = \left(\frac{2^2q^4}{2^{\frac{3}{2}}q^{\frac{3}{2}}}\right)^2 = \left(2^{2-\frac{3}{2}}q^{4-\frac{3}{2}}\right)^2 = 2^{\frac{1}{2}\cdot 2}q^{\frac{5}{2}\cdot 2} = 2q^5$$

Thus:

$$\sqrt[6]{p^5}\;\sqrt[35]{\frac{\sqrt[6]{3}\left(\left((3p)^{-\frac{1}{2}}\right)^3\right)^{-\frac{1}{3}}}{p^{-6}\sqrt[3]{(3p)^2}}} + \left(\frac{(2q)^{\frac{3}{2}}}{4q^4}\right)^{-2} = \boxed{p + 2q^5}$$

3. Substituting the values:

$$p + 2q^5 = 60 + 2\cdot 2^5 = \boxed{124}$$

Exercise 2.16

$$\left(\sqrt[3]{4}\;\sqrt[3]{\frac{16}{\sqrt{\frac{y^{-2}}{z^3}}}}\sqrt{\frac{\sqrt[3]{y}\sqrt[4]{z^3}}{16y^{0.75}}}\right)^8 = \left(\sqrt[3]{4}\left(\frac{16}{\left(\frac{1}{y^2z^3}\right)^{\frac{1}{2}}}\right)^{\frac{1}{3}}\left(\frac{y^{\frac{1}{3}}z^{\frac{3}{4}}}{16y^{\frac{3}{4}}}\right)^{\frac{1}{2}}\right)^8$$

$$= \left(4^{\frac{1}{3}}16^{\frac{1}{3}}y^{\frac{1}{3}}z^{\frac{1}{2}}y^{\frac{1}{6}}z^{\frac{3}{8}}16^{-\frac{1}{2}}y^{-\frac{3}{8}}\right)^8 = \left(4^{\frac{1}{3}}4^{\frac{2}{3}}4^{-\frac{2}{2}}y^{\frac{1}{3}+\frac{1}{6}-\frac{3}{8}}z^{\frac{1}{2}+\frac{3}{8}}\right)^8$$

$$= \left(4^{\frac{1}{3}+\frac{2}{3}-1}y^{\frac{1}{3}+\frac{1}{6}-\frac{3}{8}}z^{\frac{1}{2}+\frac{3}{8}}\right)^8 = \left(4^0y^{\frac{1}{8}}z^{\frac{7}{8}}\right)^8 = \boxed{yz^7}$$

Substituting the values:

$$yz^7 = \frac{1}{128}\cdot 2^7 = \boxed{1}$$

Exercise 2.17

1.

$$\left(\frac{\sqrt[6]{\frac{D^{-3}}{D^{1.25}\sqrt[4]{D^{-3}}}}}{D^4\sqrt[4]{\frac{D^{3.5}}{D^{-\frac{3}{2}}\sqrt{D^3}\sqrt[3]{D^{-2}}}}}\right)^{-\frac{4}{9}} = \left(\frac{\left(\frac{D^{-3}}{D^{\frac{5}{4}}D^{-\frac{3}{4}}}\right)^{\frac{1}{6}}}{D^4\left(\frac{D^{\frac{7}{2}}}{D^{-\frac{3}{2}}D^{\frac{3}{2}}D^{-\frac{2}{3}}}\right)^{\frac{1}{4}}}\right)^{-\frac{4}{9}} = \left(\frac{\left(\frac{D^{-3}}{D^{\left(\frac{5}{4}-\frac{3}{4}\right)}}\right)^{\frac{1}{6}}}{D^4\left(\frac{D^{\frac{7}{2}}}{D^{\left(-\frac{3}{2}+\frac{3}{2}-\frac{2}{3}\right)}}\right)^{\frac{1}{4}}}\right)^{-\frac{4}{9}}$$

$$= \left(\frac{\left(D^{\left(-3-\frac{5}{4}+\frac{3}{4}\right)}\right)^{\frac{1}{6}}}{D^4\left(D^{\left(\frac{7}{2}+\frac{3}{2}-\frac{3}{2}+\frac{2}{3}\right)}\right)^{\frac{1}{4}}}\right)^{-\frac{4}{9}} = \left(\frac{\left(D^{-\frac{7}{2}}\right)^{\frac{1}{6}}}{D^4\left(D^{\frac{25}{6}}\right)^{\frac{1}{4}}}\right)^{-\frac{4}{9}} = \left(\frac{D^{-\frac{7}{2}\cdot\frac{1}{6}}}{D^4 D^{\frac{25}{6}\cdot\frac{1}{4}}}\right)^{-\frac{4}{9}}$$

$$= \left(\frac{D^{-\frac{7}{12}}}{D^{4+\frac{25}{24}}}\right)^{-\frac{4}{9}} = \left(D^{-\frac{7}{12}-4-\frac{25}{24}}\right)^{-\frac{4}{9}} = \left(D^{-\frac{45}{8}}\right)^{-\frac{4}{9}} = D^{-\frac{45}{8}\cdot\left(-\frac{4}{9}\right)} = D^{\frac{5}{2}} = \sqrt{D^5}$$

2.

$$\left(\sqrt[7]{\sqrt[3]{\sqrt[4]{P^5}}}\right)^{\frac{84}{5}} = \left(\left(P^5\right)^{\frac{1}{4}\cdot\frac{1}{3}\cdot\frac{1}{7}}\right)^{\frac{84}{5}} = P^{5\cdot\frac{1}{4}\cdot\frac{1}{3}\cdot\frac{1}{7}\cdot\frac{84}{5}} = P$$

Thus:

$$\left(\frac{\sqrt[6]{\frac{D^{-3}}{D^{1.25}\sqrt[4]{D^{-3}}}}}{D^4\sqrt[4]{\frac{D^{3.5}}{D^{-\frac{3}{2}}\sqrt{D^3}\sqrt[3]{D^{-2}}}}}\right)^{-\frac{4}{9}} + \left(\sqrt[7]{\sqrt[3]{\sqrt[4]{P^5}}}\right)^{16.8} = \boxed{\sqrt{D^5} + P}$$

3. Substituting the values:

$$\sqrt{D^5} + P = \left(\sqrt{4}\right)^5 + 5 = 2^5 + 5 = \boxed{37}$$

Exercise 2.18

1.

$$\left(\sqrt{\sqrt[3]{\sqrt[4]{\sqrt[5]{\sqrt[6]{\alpha}}}}}\right)^{5!} = \left(\sqrt{\sqrt[3]{\sqrt[4]{\sqrt[5]{\sqrt[6]{\alpha}}}}}\right)^{1\cdot2\cdot3\cdot4\cdot5} = \left(\alpha^{\frac{1}{6}\cdot\frac{1}{5}\cdot\frac{1}{4}\cdot\frac{1}{3}\cdot\frac{1}{2}}\right)^{2\cdot3\cdot4\cdot5} = \alpha^{\frac{2\cdot3\cdot4\cdot5}{2\cdot3\cdot4\cdot5\cdot6}} = \sqrt[6]{\alpha}$$

2.

$$\sqrt[6!]{\left(\left(\left(\left(\left((\beta^2)^3\right)^4\right)^5\right)^6\right)^7\right)} = \left(\beta^{2\cdot3\cdot4\cdot5\cdot6\cdot7}\right)^{\frac{1}{1\cdot2\cdot3\cdot4\cdot5\cdot6}} = \beta^{\frac{2\cdot3\cdot4\cdot5\cdot6\cdot7}{2\cdot3\cdot4\cdot5\cdot6}} = \beta^7$$

Thus:

$$\left(\sqrt{\sqrt[3]{\sqrt[4]{\sqrt[5]{\sqrt[6]{\alpha}}}}}\right)^{5!} + \sqrt[6!]{\left(\left(\left(\left(\left((\beta^2)^3\right)^4\right)^5\right)^6\right)^7\right)} = \boxed{\sqrt[6]{\alpha} + \beta^7}$$

3. Substituting the values:

$$\sqrt[6]{\alpha} + \beta^7 = \sqrt[6]{64} + 2^7 = \boxed{130}$$

Exercise 2.19

1.

$$\sqrt[14]{rr^2r^3r^4r^5r^6} = \sqrt[14]{r^{1+2+3+4+5+6}} = \sqrt[14]{r^{21}} = r^{\frac{21}{14}} = r^{\frac{3}{2}} = \sqrt{r^3}$$

2.

$$t^{-1}t^2t^{-3}t^4t^{-5}t^6 = t^{-1+2-3+4-5+6} = t^3$$

Thus:

$$\sqrt[14]{rr^2r^3r^4r^5r^6} + t^{-1}t^2t^{-3}t^4t^{-5}t^6 = \boxed{\sqrt{r^3} + t^3}$$

Substituting the values:

$$\sqrt{r^3} + t^3 = \sqrt{9^3} + 3^3 = \left(\sqrt{9}\right)^3 + 3^3 = \boxed{54}$$

Exercise 2.20

(a)
1.

$$\left(\sqrt[n]{x_1}\sqrt[n]{x_2}\sqrt[n]{x_3}\cdots\sqrt[n]{x_n}\right)^n = \left(x_1^{\frac{1}{n}}x_2^{\frac{1}{n}}x_3^{\frac{1}{n}}\cdots x_n^{\frac{1}{n}}\right)^n = (x_1x_2x_3\cdots x_n)^{\frac{n}{n}} = x_1x_2x_3\cdots x_n$$

2.

$$\left(\left(\left(\left(\Delta^{-x_1}\right)^{-x_2}\right)^{-x_3}\cdots\right)^{-x_n} = \Delta^{(-x_1)(-x_2)(-x_3)\cdots(-x_n)} = \Delta^{(-1)^n x_1 x_2 x_3 \cdots x_n}\right)$$

3.

$$\left(\frac{x_1 x_2 x_3 \cdots x_n}{\Delta^{(-1)^n} x_1 x_2 x_3 \cdots x_n} \right)^{\frac{1}{x_1 x_2 x_3 \cdots x_n}} = \frac{(x_1 x_2 x_3 \cdots x_n)^{\frac{1}{x_1 x_2 x_3 \cdots x_n}}}{\left(\Delta^{(-1)^n} x_1 x_2 x_3 \cdots x_n \right)^{\frac{1}{x_1 x_2 x_3 \cdots x_n}}} = \boxed{\frac{(x_1 x_2 x_3 \cdots x_n)^{\frac{1}{x_1 x_2 x_3 \cdots x_n}}}{\Delta^{(-1)^n}}}$$

(b)

1.

$$\frac{(x_1 x_2 x_3 \cdots x_n)^{\frac{1}{x_1 x_2 x_3 \cdots x_n}}}{\Delta^{(-1)^n}} = \frac{(1 \cdot 2 \cdot 3 \cdots \cdots n)^{\frac{1}{1 \cdot 2 \cdot 3 \cdots \cdots n}}}{\Delta^{(-1)^n}} = \frac{(n!)^{\frac{1}{n!}}}{\Delta^{(-1)^n}}$$

2.

$$\left(\frac{(n!)^{\frac{1}{n!}}}{\Delta^{(-1)^n}} \right)^{n!} = \frac{(n!)^{\frac{n!}{n!}}}{\Delta^{(-1)^n n!}} = \boxed{\frac{n!}{\Delta^{(-1)^n n!}}}$$

(c)

Evaluating for $\Delta = 1$ and $n = 4$:

$$\frac{4!}{1^{(-1)^4 4!}} = \frac{1 \cdot 2 \cdot 3 \cdot 4}{1} = \boxed{24}$$

(d)

Evaluating for $\Delta = 2$ and $n = 3$:

$$\frac{3!}{2^{(-1)^3 3!}} = \frac{1 \cdot 2 \cdot 3}{\frac{1}{2^6}} = 6 \cdot 64 = \boxed{384}$$

Solutions for Chapter 3 Exercises

Exercise 3.1

(a)
$$3(w + 2) = 3w + 3 \cdot 2 = \boxed{3w + 6}$$

(b)
$$2(\alpha - 3) = 2\alpha + 2(-3) = \boxed{2\alpha - 6}$$

(c)
$$4(K + 1) = 4K + 4 \cdot 1 = \boxed{4K + 4}$$

(d)
$$(\mu - 2)5 = 5\mu + 5(-2) = \boxed{5\mu - 10}$$

(e)
$$-(T + 4) = (-1)T + (-1)4 = \boxed{-T - 4}$$

(f)
$$(3z - 3)(-2) = (-2)3z + (-2)(-3) = \boxed{-6z + 6}$$

(g)
$$3(-3\rho + \beta - 4) = 3(-3\rho) + 3\beta + 3(-4) = \boxed{-9\rho + 3\beta - 12}$$

(h)
$$-(-2Y - r - 1) = (-1)(-2Y) + (-1)(-r) + (-1)(-1) = \boxed{2Y + r + 1}$$

(i)
$$(-4k - 2 - t)(-3) = (-3)(-4k) + (-3)(-2) + (-3)t = \boxed{12k + 6 + 3t}$$

(j)
$$h(-5h^2 + h - 4) = h(-5h^2) + hh + h(-4) = \boxed{-5h^3 + h^2 - 4h}$$

(k)

$$-7X\left(-2X - \frac{1}{X} - 1\right)$$

$$= -7X(-2X) + (-7X)\left(-\frac{1}{X}\right) + (-7X)(-1) = \boxed{14X^2 + 7 + 7X}$$

(l)

$$\left(-\frac{7}{D} - \frac{2}{D^2} - D\right)(-3D)$$

$$= (-3D)\left(-\frac{7}{D}\right) + (-3D)\left(-\frac{2}{D^2}\right) + (-3D)(-D) = \boxed{21 + \frac{6}{D} + 3D^2}$$

Exercise 3.2

(a)

$$3b^2(4b^3 + 5b^4) = 3b^2(4b^3) + 3b^2(5b^4) = \boxed{12b^5 + 15b^6}$$

(b)

$$4x^3(2x^2 + 3x^4) = 4x^3(2x^2) + 4x^3(3x^4) = \boxed{8x^5 + 12x^7}$$

(c)

$$5D^4(3D^2 - 4D^3) = 5D^4(3D^2) + 5D^4(-4D^3) = \boxed{15D^6 - 20D^7}$$

(d)

$$(3\mu^2 + 4\mu^4)2\mu^3 = 2\mu^3(3\mu^2) + 2\mu^3(4\mu^4) = \boxed{6\mu^5 + 8\mu^7}$$

(e)

$$(2\nu^3 - 5\nu^2)3\nu^4 = 3\nu^4(2\nu^3) + 3\nu^4(-5\nu^2) = \boxed{6\nu^7 - 15\nu^6}$$

(f)

$$-6E^2(2E^3 - 3E^4) = -6E^2(2E^3) + (-6E^2)(-3E^4) = \boxed{-12E^5 + 18E^6}$$

(g)

$$-2z^4(-4z^2 - 5z^3) = -2z^4(-4z^2) + (-2z^4)(-5z^3) = \boxed{8z^6 + 10z^7}$$

(h)

$$(-3G^3 + 2G^2)(-3G^4) = -3G^4(-3G^3) + (-3G^4)(2G^2) = \boxed{9G^7 - 6G^6}$$

(i)

$$-5\lambda^3(-4\lambda^2 - 3\lambda^4) = -5\lambda^3(-4\lambda^2) + (-5\lambda^3)(-3\lambda^4) = \boxed{20\lambda^5 + 15\lambda^7}$$

(j)

$$-4H^3(-2K^4 - 5K^2) = -4H^3(-2K^4) + (-4H^3)(-5K^2) = \boxed{8H^3K^4 + 20H^3K^2}$$

(k)

$$(3\alpha^3 + 4\alpha^2)(-2\phi^4) = -2\phi^4(3\alpha^3) + (-2\phi^4)(4\alpha^2) = \boxed{-6\phi^4\alpha^3 - 8\phi^4\alpha^2}$$

(l)

$$-3I^2(-5N^4 - 2M^3) = -3I^2(-5N^4) + (-3I^2)(-2M^3) = \boxed{15I^2N^4 + 6I^2M^3}$$

Exercise 3.3

(a)

$$\sqrt[4]{s^5}\left(\sqrt[6]{s^5} + \sqrt[8]{s^7}\right) = s^{\frac{5}{4}}\left(s^{\frac{5}{6}} + s^{\frac{7}{8}}\right) = s^{\frac{5}{4}}s^{\frac{5}{6}} + s^{\frac{5}{4}}s^{\frac{7}{8}} = s^{\frac{25}{12}} + s^{\frac{17}{8}} = \boxed{\sqrt[12]{s^{25}} + \sqrt[8]{s^{17}}}$$

(b)

$$\sqrt[5]{t^6}\left(\sqrt[4]{t^3} + \sqrt[7]{t^4}\right) = t^{\frac{6}{5}}\left(t^{\frac{3}{4}} + t^{\frac{4}{7}}\right) = t^{\frac{6}{5}}t^{\frac{3}{4}} + t^{\frac{6}{5}}t^{\frac{4}{7}} = t^{\frac{39}{20}} + t^{\frac{62}{35}} = \boxed{\sqrt[20]{t^{39}} + \sqrt[35]{t^{62}}}$$

(c)

$$\left(\sqrt[3]{u^4} + \sqrt[9]{u^5}\right)\sqrt[7]{u^3} = \left(u^{\frac{4}{3}} + u^{\frac{5}{9}}\right)u^{\frac{3}{7}} = u^{\frac{4}{3}}u^{\frac{3}{7}} + u^{\frac{5}{9}}u^{\frac{3}{7}} = u^{\frac{37}{21}} + u^{\frac{62}{63}} = \boxed{\sqrt[21]{u^{37}} + \sqrt[63]{u^{62}}}$$

(d)

$$\sqrt[5]{v^7}\left(\sqrt[5]{v^3} - \sqrt[4]{v^5}\right) = -v^{\frac{7}{5}}\left(v^{\frac{3}{5}} - v^{\frac{5}{4}}\right) = -v^{\frac{7}{5}}v^{\frac{3}{5}} + v^{\frac{7}{5}}v^{\frac{5}{4}} = -v^2 + v^{\frac{53}{20}} = \boxed{-v^2 + \sqrt[20]{v^{53}}}$$

(e)

$$- \sqrt[7]{w^2} \left(-\sqrt[7]{w^5} - \sqrt[3]{w^4} \right) = -w^{\frac{2}{7}} \left(-w^{\frac{5}{7}} - w^{\frac{4}{3}} \right)$$

$$= w^{\frac{2}{7}} w^{\frac{5}{7}} + w^{\frac{2}{7}} w^{\frac{4}{3}} = w^{\frac{7}{7}} + w^{\frac{34}{21}} = \boxed{w + \sqrt[21]{w^{34}}}$$

(f)

$$\left(\sqrt[5]{x^4} - \sqrt[8]{x^5} \right) \sqrt[9]{x^5} = \left(x^{\frac{4}{5}} - x^{\frac{5}{8}} \right) x^{\frac{5}{9}} = x^{\frac{4}{5}} x^{\frac{5}{9}} - x^{\frac{5}{8}} x^{\frac{5}{9}} = x^{\frac{61}{45}} - x^{\frac{85}{72}} = \boxed{\sqrt[45]{x^{61}} - \sqrt[72]{x^{85}}}$$

(g)

$$\left(-\sqrt[3]{y^5} - \sqrt[7]{y^3} \right) \sqrt[8]{y^5} = \left(-y^{\frac{5}{3}} - y^{\frac{3}{7}} \right) y^{\frac{5}{8}}$$

$$= -y^{\frac{5}{3}} y^{\frac{5}{8}} - y^{\frac{3}{7}} y^{\frac{5}{8}} = -y^{\frac{55}{24}} - y^{\frac{59}{56}} = \boxed{-\sqrt[24]{y^{55}} - \sqrt[56]{y^{59}}}$$

(h)

$$\sqrt[5]{z^8} \left(\sqrt[4]{z^4} + \sqrt[10]{z^3} + \sqrt[7]{z^6} \right) = z^{\frac{8}{5}} \left(z^{\frac{4}{4}} + z^{\frac{3}{10}} + z^{\frac{6}{7}} \right)$$

$$= z^{\frac{8}{5}} z^{\frac{4}{4}} + z^{\frac{8}{5}} z^{\frac{3}{10}} + z^{\frac{8}{5}} z^{\frac{6}{7}} = z^{\frac{13}{5}} + z^{\frac{19}{10}} + z^{\frac{86}{35}} = \boxed{\sqrt[5]{z^{13}} + \sqrt[10]{z^{19}} + \sqrt[35]{z^{86}}}$$

Exercise 3.4

(a)

$$8p + 24 = 8p + (8)3 = \boxed{8(p + 3)}$$

(b)

$$5\beta - 15 = 5\beta + (5)(-3) = \boxed{5(\beta - 3)}$$

(c)

$$9A + 27 = 9A + (9)3 = \boxed{9(A + 3)}$$

(d)

$$6\delta - 18 = 6\delta + (6)(-3) = \boxed{6(\delta - 3)}$$

(e)

$$-3E - 9 = -3E + (-3)3 = \boxed{-3(E + 3)}$$

(f)

$$-f - 2 = -1f + (-1)2 = \boxed{-(f + 2)}$$

(g)
$$20\gamma + 30\eta - 50 = 10(2\gamma) + 10(3\eta) + 10(-5) = \boxed{10(2\gamma + 3\eta - 5)}$$

(h)
$$-9H + 3\theta + 12 = (-3)(3H) + (-3)(-\theta) + (-3)(-4) = \boxed{-3(3H - \theta - 4)}$$

(i)
$$12i + 16j - 8 = (4)3i + (4)4j + (4)(-2) = \boxed{4(3i + 4j - 2)}$$

(j)
$$-4k^3 + 2k^2 - 6k = (-2k)2k^2 + (-2k)k + (-2k)3 = \boxed{-2k(2k^2 - k + 3)}$$

(k)
$$-15L^4 - 5L^3 - 20L^2 = (-5L^2)3L^2 + (-5L^2)L + (-5L^2)4 = \boxed{-5L^2(3L^2 + L + 4)}$$

(l)
$$-10M^5 - 5N^4 - 30N^2 = (-5)2M^5 + (-5)N^4 + (-5)6N^2 = \boxed{-5(2M^5 + N^4 + 6N^2)}$$

Exercise 3.5

(a)
$$14a^4b + 21a^5b = (7a^4b)2 + (7a^4b)3a = \boxed{7a^4b(2 + 3a)}$$

(b)
$$10bc^6 + 15bc^8 = (5bc^6)2 + (5bc^6)3c^2 = \boxed{5bc^6(2 + 3c^2)}$$

(c)
$$18p^5s - 12sp^6 = (6p^5s)3 + (6p^5s)(-2p) = \boxed{6p^5s(3 - 2p)}$$

(d)
$$9qr^6 + 27qr^7 = (9qr^6)1 + (9qr^6)3r = \boxed{9qr^6(1 + 3r)}$$

(e)
$$16s^8p^3 - 8p^3s^7 = (8s^7p^3)2s + (8s^7p^3)(-1) = \boxed{8s^7p^3(2s - 1)}$$

(f)
$$-20t^4k^5 + 30k^5t^5 = (-10t^4k^5)2 + (-10t^4k^5)(-3t) = \boxed{-10t^4k^5(2 - 3t)}$$

(g)

$$-28u^3v^2 + 14u^4v^2 = (-14u^3v^2)2 + (-14u^3v^2)(-u) = \boxed{-14u^3v^2(2-u)}$$

(h)

$$25w^4v^6 - 10v^5w = (5v^5w)5vw^3 + (5v^5w)(-2) = \boxed{5v^5w(5vw^3 - 2)}$$

(i)

$$-21w^7z^4 + 7w^9z^5 = (-7w^7z^4)3 + (-7w^7z^4)(-w^2z) = \boxed{-7w^7z^4(3 - w^2z)}$$

(j)

$$12y^4x^3 - 18x^3y^5 - 6y^4x^5$$

$$= (6x^3y^4)2 + (6x^3y^4)(-3y) + (6x^3y^4)(-x^2) = \boxed{6x^3y^4(2 - 3y - x^2)}$$

(k)

$$-8z^4\theta^3 - 16\theta^2z^4 - 8z^4\theta^5 = (-8z^4\theta^2)\theta + (-8z^4\theta^2)2 + (-8z^4\theta^2)\theta^3 = \boxed{-8z^4\theta^2(\theta + 2 + \theta^3)}$$

(l)

$$21\beta^4\alpha^2 - 7\alpha^2\beta^3 - 7\beta^9\alpha^2$$

$$= (7\alpha^2\beta^3)3\beta + (7\alpha^2\beta^3)(-1) + (7\alpha^2\beta^3)(-\beta^6) = \boxed{7\alpha^2\beta^3(3\beta - 1 - \beta^6)}$$

Exercise 3.6

(a)

$$1 + t = -t^4\left(-\frac{1}{t^4}\right) - t^4\left(-\frac{t}{t^4}\right) = -t^4\left(-\frac{1}{t^4}\right) - t^4\left(\frac{1}{t^3}\right) = \boxed{-t^4\left(-\frac{1}{t^4} - \frac{1}{t^3}\right)}$$

(b)

$$t^3 - t^2 = -t^4\left(-\frac{t^3}{t^4}\right) - t^4\left(\frac{t^2}{t^4}\right) = -t^4\left(-\frac{t}{t}\right) - t^4\left(\frac{t}{t^2}\right) = \boxed{-t^4\left(-\frac{1}{t} + \frac{1}{t^2}\right)}$$

(c)

$$-t^9 + t^3 = -t^4\left(\frac{t^9}{t^4}\right) - t^4\left(-\frac{t^3}{t^4}\right) = -t^4t^5 - t^4\left(-\frac{1}{t}\right) = \boxed{-t^4\left(t^5 - \frac{1}{t}\right)}$$

(d)

$$-2 - t - t^2 = -t^4 \left(\frac{2}{t^4} \right) - t^4 \left(\frac{t}{t^4} \right) - t^4 \left(\frac{t^2}{t^4} \right) = \boxed{-t^4 \left(\frac{2}{t^4} + \frac{1}{t^3} + \frac{1}{t^2} \right)}$$

(e)

$$t^8 - t^5 + t^4 - 1 = -t^4 \left(-\frac{t^8}{t^4} \right) - t^4 \left(\frac{t^5}{t^4} \right) - t^4 \left(-\frac{t^4}{t^4} \right) - t^4 \left(\frac{1}{t^4} \right)$$

$$= \boxed{-t^4 \left(-t^4 + t - 1 + \frac{1}{t^4} \right)}$$

(f)

$$t^6 - 3t^4 + 2t^2 = -t^4 \left(-\frac{t^6}{t^4} \right) - t^4 \left(\frac{3t^4}{t^4} \right) - t^4 \left(-\frac{2t^2}{t^4} \right) = \boxed{-t^4 \left(-t^2 + 3 - \frac{2}{t^2} \right)}$$

(g)

$$2t^{\frac{8}{5}} - t^7 + t^{\frac{3}{4}} - 2 = -t^4 \left(-\frac{2t^{\frac{8}{5}}}{t^4} \right) - t^4 \left(\frac{t^7}{t^4} \right) - t^4 \left(-\frac{t^{\frac{3}{4}}}{t^4} \right) - t^4 \left(\frac{2}{t^4} \right)$$

$$= -t^4 \left(-2t^{-\frac{12}{5}} + t^3 - t^{-\frac{13}{4}} + \frac{2}{t^4} \right) = \boxed{-t^4 \left(-2\sqrt[5]{\frac{1}{t^{12}}} + t^3 - \sqrt[4]{\frac{1}{t^{13}}} + \frac{2}{t^4} \right)}$$

(h)

$$-t^{12} - t^{\frac{9}{4}} + t^{\frac{2}{5}} - t = -t^4 \left(\frac{t^{12}}{t^4} \right) - t^4 \left(\frac{t^{\frac{9}{4}}}{t^4} \right) - t^4 \left(-\frac{t^{\frac{2}{5}}}{t^4} \right) - t^4 \left(\frac{t}{t^4} \right)$$

$$= -t^4 \left(t^8 + t^{-\frac{7}{4}} - t^{-\frac{18}{5}} + t^{-3} \right) = \boxed{-t^4 \left(t^8 + \sqrt[4]{\frac{1}{t^7}} - \sqrt[5]{\frac{1}{t^{18}}} + \frac{1}{t^3} \right)}$$

(i)

$$-t^4 - t^{-\frac{5}{2}} + t^{-\frac{2}{3}} = -t^4 \left(\frac{t^4}{t^4} \right) - t^4 \left(\frac{t^{-\frac{5}{2}}}{t^4} \right) - t^4 \left(-\frac{t^{-\frac{2}{3}}}{t^4} \right)$$

$$= -t^4 \left(1 + t^{-\frac{13}{2}} - t^{-\frac{14}{3}} \right) = \boxed{-t^4 \left(1 + \sqrt{\frac{1}{t^{13}}} - \sqrt[3]{\frac{1}{t^{14}}} \right)}$$

Exercise 3.7

(a)

$$\sqrt[3]{x^{10}} + \sqrt[3]{x^8} = x^{\frac{10}{3}} + x^{\frac{8}{3}} = x^{\frac{7}{3}} x^{\frac{3}{3}} + x^{\frac{7}{3}} x^{\frac{1}{3}} = x^{\frac{7}{3}} \left(x + x^{\frac{1}{3}} \right) = \boxed{\sqrt[3]{x^7} \left(x + \sqrt[3]{x} \right)}$$

(b)

$$-\sqrt[3]{x^{11}} - \sqrt[4]{x^{15}} = -x^{\frac{11}{3}} - x^{\frac{15}{4}} = x^{\frac{7}{3}} \left(-x^{\frac{4}{3}} \right) + x^{\frac{7}{3}} \left(-x^{\frac{17}{12}} \right)$$

$$= x^{\frac{7}{3}} \left(-x^{\frac{4}{3}} - x^{\frac{17}{12}} \right) = \boxed{\sqrt[3]{x^7} \left(-\sqrt[3]{x^4} - \sqrt[12]{x^{17}} \right)}$$

(c)

$$\sqrt[5]{x^{21}} - \sqrt{x^{11}} = x^{\frac{21}{5}} - x^{\frac{11}{2}} = x^{\frac{7}{3}} x^{\frac{28}{15}} - x^{\frac{7}{3}} x^{\frac{19}{6}} = x^{\frac{7}{3}} \left(x^{\frac{28}{15}} - x^{\frac{19}{6}} \right) = \boxed{\sqrt[3]{x^7} \left(\sqrt[15]{x^{28}} - \sqrt[6]{x^{19}} \right)}$$

(d)

$$\sqrt[15]{x^{14}} + \sqrt[6]{x^7} = x^{\frac{14}{15}} + x^{\frac{7}{6}} = x^{\frac{7}{3}} x^{-\frac{7}{5}} + x^{\frac{7}{3}} x^{-\frac{7}{6}} = x^{\frac{7}{3}} \left(x^{-\frac{7}{5}} + x^{-\frac{7}{6}} \right) = \boxed{\sqrt[3]{x^7} \left(\sqrt[5]{\frac{1}{x^7}} + \sqrt[6]{\frac{1}{x^7}} \right)}$$

(e)

$$-\sqrt[3]{x^{14}} - \sqrt[15]{x^{28}} = -x^{\frac{14}{3}} - x^{\frac{28}{15}} = -x^{\frac{7}{3}} x^{\frac{7}{3}} - x^{\frac{7}{3}} x^{-\frac{7}{15}}$$

$$= x^{\frac{7}{3}} \left(-x^{\frac{7}{3}} - x^{-\frac{7}{15}} \right) = \boxed{\sqrt[3]{x^7} \left(-\sqrt[3]{x^7} - \sqrt[15]{\frac{1}{x^7}} \right)}$$

(f)

$$-\sqrt[9]{x^{14}} - \sqrt[4]{x^{21}} = -x^{\frac{14}{9}} - x^{\frac{21}{4}} = -x^{\frac{7}{3}} x^{-\frac{7}{9}} - x^{\frac{7}{3}} x^{\frac{35}{12}}$$

$$= x^{\frac{7}{3}} \left(-x^{-\frac{7}{9}} - x^{\frac{35}{12}} \right) = \boxed{\sqrt[3]{x^7} \left(-\sqrt[9]{\frac{1}{x^7}} - \sqrt[12]{x^{35}} \right)}$$

(g)

$$\sqrt[4]{x^5} + \sqrt{x^5} = x^{\frac{5}{4}} + x^{\frac{5}{2}} = x^{\frac{7}{3}} x^{-\frac{13}{12}} + x^{\frac{7}{3}} x^{\frac{1}{6}} = x^{\frac{7}{3}} \left(x^{-\frac{13}{12}} + x^{\frac{1}{6}} \right) = \boxed{\sqrt[3]{x^7} \left(\sqrt[12]{\frac{1}{x^{13}}} + \sqrt[6]{x} \right)}$$

(h)

$$\sqrt[5]{x^3} + \sqrt[4]{x^7} = x^{\frac{3}{5}} + x^{\frac{7}{4}} = x^{\frac{7}{3}}x^{-\frac{26}{15}} + x^{\frac{7}{3}}x^{-\frac{7}{12}}$$

$$= x^{\frac{7}{3}}\left(x^{-\frac{26}{15}} + x^{-\frac{7}{12}}\right) = \boxed{\sqrt[3]{x^7}\left(\sqrt[15]{\frac{1}{x^{26}}} + \sqrt[12]{\frac{1}{x^7}}\right)}$$

(i)

$$-\sqrt{x^5} - \sqrt[8]{x^3} = -x^{\frac{5}{2}} - x^{\frac{3}{8}} = -x^{\frac{7}{3}}x^{\frac{1}{6}} - x^{\frac{7}{3}}x^{-\frac{47}{24}}$$

$$= x^{\frac{7}{3}}\left(-x^{\frac{1}{6}} - x^{-\frac{47}{24}}\right) = \boxed{\sqrt[3]{x^7}\left(-\sqrt[6]{x} - \sqrt[24]{\frac{1}{x^{47}}}\right)}$$

Exercise 3.8

(a)

$$3a + 5a - 2a = (3 + 5 - 2)a = \boxed{6a}$$

(b)

$$-7x + 3x^2 - 4x^2 + 2x = (3 - 4)x^2 + (-7 + 2)x = \boxed{-x^2 - 5x}$$

(c)

$$4\beta + 2\gamma - 3\beta + \gamma = (4 - 3)\beta + (2 + 1)\gamma = \boxed{\beta + 3\gamma}$$

(d)

$$5A^2 - 3A^3 + 2A^3 - 4A^2 = (-3 + 2)A^3 + (5 - 4)A^2 = \boxed{-A^3 + A^2}$$

(e)

$$3\sqrt{p} - 4\sqrt{p^3} + 2\sqrt{p^3} + \sqrt{p} = (3 + 1)\sqrt{p} + (-4 + 2)\sqrt{p^3} = \boxed{4\sqrt{p} - 2\sqrt{p^3}}$$

(f)

$$-2\sqrt[3]{y^5} + 5\sqrt[3]{y^5} - 3\sqrt[3]{y^5} = (-2 + 5 - 3)\sqrt[3]{y^5} = \boxed{0}$$

(g)

$$3\alpha^2\beta - 2\alpha\beta^2 + \alpha^2\beta - 4\alpha\beta^2 = (3 + 1)\alpha^2\beta + (-2 - 4)\alpha\beta^2 = \boxed{4\alpha^2\beta - 6\alpha\beta^2}$$

(h)

$$2\sqrt{x} + 3\sqrt[3]{x^2} - \sqrt{x} - \sqrt[3]{x^2} = (2 - 1)\sqrt{x} + (3 - 1)\sqrt[3]{x^2} = \boxed{\sqrt{x} + 2\sqrt[3]{x^2}}$$

(i)

$$-\frac{1}{2}\sqrt[4]{m^3} - \frac{3}{4}\sqrt[4]{m^3} + \frac{1}{4}\sqrt[4]{m^3} = \left(-\frac{1}{2} - \frac{3}{4} + \frac{1}{4}\right)\sqrt[4]{m^3} = \boxed{-\sqrt[4]{m^3}}$$

(j)

$$4V^2 - \frac{1}{2}V^3 + \frac{3}{2}V^3 + 2V^2 = \left(\frac{3}{2} - \frac{1}{2}\right)V^3 + (4+2)V^2 = \boxed{V^3 + 6V^2}$$

(k)

$$-\frac{1}{3}\sqrt[5]{n^4} + \frac{2}{3}\sqrt[5]{n^4} - \frac{1}{3}\sqrt[5]{n^4} = \left(-\frac{1}{3} + \frac{2}{3} - \frac{1}{3}\right)\sqrt[5]{n^4} = \boxed{0}$$

(l)

$$3\omega^2 - 2\omega^3 + \omega^2 - 4\omega^3 = (-2-4)\omega^3 + (3+1)\omega^2 = \boxed{-6\omega^3 + 4\omega^2}$$

(m)

$$5\sqrt[3]{r^5} + 2\sqrt[5]{r^3} - 3\sqrt[3]{r^5} + \sqrt[3]{r^5} = (5-3+1)\sqrt[3]{r^5} + 2\sqrt[5]{r^3} = \boxed{3\sqrt[3]{r^5} + 2\sqrt[5]{r^3}}$$

(n)

$$-3\theta^3 + 4\theta^2 - 2\theta^3 + 3\theta^2 = (-3-2)\theta^3 + (4+3)\theta^2 = \boxed{-5\theta^3 + 7\theta^2}$$

(o)

$$2\sqrt[4]{s^6} - \sqrt[4]{s^6} + \frac{1}{2}\sqrt[4]{s^6} = \left(2 - 1 + \frac{1}{2}\right)\sqrt[4]{s^6} = \boxed{\frac{3}{2}\sqrt[4]{s^6}}$$

(p)

$$-\frac{3}{4}\psi^4 + \frac{1}{4}\psi^4 - \frac{1}{2}\psi^4 + \frac{3}{4}\psi^4 = \left(-\frac{3}{4} + \frac{1}{4} - \frac{1}{2} + \frac{3}{4}\right)\psi^4 = \boxed{-\frac{1}{4}\psi^4}$$

(q)

$$4\eta^3\phi^2 - 2\eta^3\phi^2 + \eta^3\phi^2 - 3\eta^3\phi^2 = (4-2+1-3)\eta^3\phi^2 = \boxed{0}$$

(r)

$$-5\kappa\lambda + 3\kappa - 4\kappa\lambda - \kappa\lambda = (-5-4-1)\kappa\lambda + 3\kappa = \boxed{-10\kappa\lambda + 3\kappa}$$

Exercise 3.9

(a)

$$-\left(3A^2B + 3\sqrt{C} + 2\sqrt[3]{D^4} - 5\sqrt[3]{D^4}\right) + A^2B - 2\sqrt{C} - 3A^2B + 4\sqrt{C} + 2\sqrt[3]{D^4} + \sqrt{C}$$
$$= -3A^2B - 3\sqrt{C} - 2\sqrt[3]{D^4} + 5\sqrt[3]{D^4} + A^2B - 2\sqrt{C} + 4\sqrt{C} + 2\sqrt[3]{D^4} + \sqrt{C}$$
$$= (-3+1-3)A^2B + (-3-2+4+1)\sqrt{C} + (-2+5+2)\sqrt[3]{D^4}$$
$$= \boxed{-5A^2B + 5\sqrt[3]{D^4}}$$

(b)

$$-2\alpha\beta + \delta^3 - \alpha\beta - \delta^3 - (3\sqrt[3]{\gamma^2} + 5\sqrt[3]{\gamma^2} - 4\delta^3 + 2\delta^3 + 3\alpha\beta + 3\delta^3) - 2\sqrt[3]{\gamma^2}$$

$$= -2\alpha\beta + \delta^3 - \alpha\beta - \delta^3 - 3\sqrt[3]{\gamma^2} - 5\sqrt[3]{\gamma^2} + 4\delta^3 - 2\delta^3 - 3\alpha\beta - 3\delta^3 - 2\sqrt[3]{\gamma^2}$$

$$= (-2 - 1 - 3)\alpha\beta + (1 - 1 + 4 - 2 - 3)\delta^3 + (-3 - 5 - 2)\sqrt[3]{\gamma^2}$$

$$= \boxed{-6\alpha\beta - \delta^3 - 10\sqrt[3]{\gamma^2}}$$

(c)

$$5\sqrt[3]{W^2} - (2X^2Y^3 + 3\sqrt[4]{Z} + 2\sqrt[3]{W^2} - 2\sqrt[4]{Z} - 3\sqrt[3]{W^2} + \sqrt[4]{Z} + X^2Y^3 - \sqrt[3]{W^2})$$

$$= 5\sqrt[3]{W^2} - 2X^2Y^3 - 3\sqrt[4]{Z} - 2\sqrt[3]{W^2} + 2\sqrt[4]{Z} + 3\sqrt[3]{W^2} - \sqrt[4]{Z} - X^2Y^3 + \sqrt[3]{W^2}$$

$$= (5 - 2 + 3 + 1)\sqrt[3]{W^2} + (-2 - 1)X^2Y^3 + (-3 + 2 - 1)\sqrt[4]{Z}$$

$$= \boxed{7\sqrt[3]{W^2} - 3X^2Y^3 - 2\sqrt[4]{Z}}$$

(d)

$$-(\sqrt[5]{r^4} - 2p^2q^3 + \sqrt[5]{r^4} - s^3) - (5p^2q^3 - 2s^3 - 3p^2q^3 + \sqrt[5]{r^4} + 3s^3 - 3p^2q^3)$$

$$= -\sqrt[5]{r^4} + 2p^2q^3 - \sqrt[5]{r^4} + s^3 - 5p^2q^3 + 2s^3 + 3p^2q^3 - \sqrt[5]{r^4} - 3s^3 + 3p^2q^3$$

$$= (-1 - 1 - 1)\sqrt[5]{r^4} + (2 - 5 + 3 + 3)p^2q^3 + (1 + 2 - 3)s^3$$

$$= \boxed{-3\sqrt[5]{r^4} + 3p^2q^3}$$

(e)

$$-(3G^2H - \sqrt{E} - 4\sqrt[4]{F^3} + G^2H + \sqrt[4]{F^3}) - (2\sqrt{E} - 3G^2H + 3\sqrt{E} + 5G^2H)$$

$$= -3G^2H + \sqrt{E} + 4\sqrt[4]{F^3} - G^2H - \sqrt[4]{F^3} - 2\sqrt{E} + 3G^2H - 3\sqrt{E} - 5G^2H$$

$$= (-3 - 1 + 3 - 5)G^2H + (1 - 2 - 3)\sqrt{E} + (4 - 1)\sqrt[4]{F^3}$$

$$= \boxed{-6G^2H - 4\sqrt{E} + 3\sqrt[4]{F^3}}$$

(f)

$$-(x^3y - 2\sqrt[3]{u^4}) - (x^3y - 2v^2w) - (4\sqrt[3]{u^4} + \sqrt[3]{u^4} + 3v^2w - x^3y - v^2w + 5x^3y)$$

$$= -x^3y + 2\sqrt[3]{u^4} - x^3y + 2v^2w - 4\sqrt[3]{u^4} - \sqrt[3]{u^4} - 3v^2w + x^3y + v^2w - 5x^3y$$

$$= (-1 - 1 + 1 - 5)x^3y + (2 - 4 - 1)\sqrt[3]{u^4} + (2 - 3 + 1)v^2w$$

$$= \boxed{-6x^3y - 3\sqrt[3]{u^4}}$$

(g)

$$-(3\Omega^3\Psi^2 - 2\sqrt[5]{N^3} + \Omega^3\Psi^2) - (3\sqrt[5]{N^3} + \sqrt[5]{N^3} - 2\sqrt[3]{M^5} + 4\Omega^3\Psi^2 - \sqrt[3]{M^5} - 3\Omega^3\Psi^2)$$
$$= -3\Omega^3\Psi^2 + 2\sqrt[5]{N^3} - \Omega^3\Psi^2 - 3\sqrt[5]{N^3} - \sqrt[5]{N^3} + 2\sqrt[3]{M^5} - 4\Omega^3\Psi^2 + \sqrt[3]{M^5} + 3\Omega^3\Psi^2$$
$$= (-3 - 1 - 4 + 3)\Omega^3\Psi^2 + (2 - 3 - 1)\sqrt[5]{N^3} + (2+1)\sqrt[3]{M^5}$$
$$= \boxed{-5\Omega^3\Psi^2 - 2\sqrt[5]{N^3} + 3\sqrt[3]{M^5}}$$

(h)

$$3\delta^3 - (\alpha\beta + 2\delta^3 - 2\alpha\beta + \delta^3) + 3\alpha\beta + 2\sqrt[3]{Q^2} - 3\sqrt[3]{Q^2} + 5\sqrt[3]{Q^2} - (\delta^3 - 4\delta^3)$$
$$= 3\delta^3 - \alpha\beta - 2\delta^3 + 2\alpha\beta - \delta^3 + 3\alpha\beta + 2\sqrt[3]{Q^2} - 3\sqrt[3]{Q^2} + 5\sqrt[3]{Q^2} - \delta^3 + 4\delta^3$$
$$= (3 - 2 - 1 - 1 + 4)\delta^3 + (-1 + 2 + 3)\alpha\beta + (2 - 3 + 5)\sqrt[3]{Q^2}$$
$$= \boxed{3\delta^3 + 4\alpha\beta + 4\sqrt[3]{Q^2}}$$

(i)

$$-(\sqrt[3]{S^2} + X^2Y^3) + 3\sqrt[3]{S^2} - (3\sqrt[4]{R} + X^2Y^3) + 5\sqrt[3]{S^2} + 2\sqrt[4]{R} - (2X^2Y^3 - 2\sqrt[3]{S^2})$$
$$= -\sqrt[3]{S^2} - X^2Y^3 + 3\sqrt[3]{S^2} - 3\sqrt[4]{R} - X^2Y^3 + 5\sqrt[3]{S^2} + 2\sqrt[4]{R} - 2X^2Y^3 + 2\sqrt[3]{S^2}$$
$$= (3 + 5 + 2 - 1)\sqrt[3]{S^2} + (-1 - 1 - 2)X^2Y^3 + (-3 + 2)\sqrt[4]{R}$$
$$= \boxed{9\sqrt[3]{S^2} - 4X^2Y^3 - \sqrt[4]{R}}$$

(j)

$$-W^3 - (3U^2V^3 + \sqrt[5]{T^4} - 2W^3) - 2U^2V^3 - (\sqrt[5]{T^4} - 3U^2V^3 + \sqrt[5]{T^4}) + 3W^3$$
$$= -W^3 - 3U^2V^3 - \sqrt[5]{T^4} + 2W^3 - 2U^2V^3 - \sqrt[5]{T^4} + 3U^2V^3 - \sqrt[5]{T^4} + 3W^3$$
$$= (-1 + 2 + 3)W^3 + (-3 - 2 + 3)U^2V^3 + (-1 - 1 - 1)\sqrt[5]{T^4}$$
$$= \boxed{4W^3 - 2U^2V^3 - 3\sqrt[5]{T^4}}$$

Exercise 3.10

(a)

$$-2x(x^3 - 3x + 1) - x^2(-3x^2 + 2) - x(x^3 + 4x - 2)$$
$$= -2x^4 + 6x^2 - 2x + 3x^4 - 2x^2 - x^4 - 4x^2 + 2x$$
$$= (-2 + 3 - 1)x^4 + (6 - 2 - 4)x^2 + (-2 + 2)x = \boxed{0}$$

(b)

$$3p(4p^2 - 5q + 1) - 2q(6p^2 + 7q) - p(12p^2 - 12pq - 15q + 3) - q(4 - 14q)$$
$$= 12p^3 - 15pq + 3p - 12p^2q - 14q^2 - 12p^3 + 12p^2q + 15pq - 3p - 4q + 14q^2$$
$$= (12 - 12)p^3 + (-12 + 12)p^2q + (-15 + 15)pq + (-14 + 14)q^2 + (3 - 3)p - 4q$$
$$= \boxed{-4q}$$

(c)

$$-5r(3r^2 + 2s - 4r^3) + 2s(4r^2 - 3r^3 + s) - r(20r^3 - 6r^2s - 15r^2 + 8rs - 10s)$$
$$= -15r^3 - 10rs + 20r^4 + 8r^2s - 6r^3s + 2s^2 - 20r^4 + 6r^3s + 15r^3 - 8r^2s + 10rs$$
$$= (20 - 20)r^4 + (-6 + 6)r^3s + (8 - 8)r^2s + (-10 + 10)rs + 2s^2 = \boxed{2s^2}$$

(d)

$$-2V(6V^2 - 4V^3) + W(3V^2 - 2W^2) - V(8V^3 - 12V^2 - 3VW) - (6V^2W - 3W^3)$$
$$= -12V^3 + 8V^4 + 3V^2W - 2W^3 - 8V^4 + 12V^3 + 3V^2W - 6V^2W + 3W^3$$
$$= (8 - 8)V^4 + (-12 + 12)V^3 + (3 + 3 - 6)V^2W + (-2 + 3)W^3 = \boxed{W^3}$$

(e)

$$-6t(8t^2 + 9t^3) + u(3t^2 - 5u^2) - t^2(-54t^2 - 48t) - u(3t^2 - 5u^2 - t)$$
$$= -48t^3 - 54t^4 + 3ut^2 - 5u^3 + 54t^4 + 48t^3 - 3ut^2 + 5u^3 + ut$$
$$= (-48 + 48)t^3 + (-54 + 54)t^4 + (3 - 3)ut^2 + (-5 + 5)u^3 + ut = \boxed{ut}$$

(f)

$$c^2(3c + c(1 + 2c^3)) - c(3c^2 + (2c^3 + 1)c^2 - c)$$
$$= 3c^3 + c^3(1 + 2c^3) - 3c^3 - (2c^3 + 1)c^3 + c^2$$
$$= 3c^3 + c^3 + 2c^6 - 3c^3 - 2c^6 - c^3 + c^2$$
$$= (3 + 1 - 3 - 1)c^3 + (2 - 2)c^6 + c^2 = \boxed{c^2}$$

(g) 1. First, let's work with the first term:

$$(((1 + n)n)n)n = ((n + n^2)n)n = (n^2 + n^3)n = n^3 + n^4$$

2. Let's now work with the second term:

$$-m(n^3(n+2+m)) = -m(n^4 + 2n^3 + mn^3) = -mn^4 - 2mn^3 - m^2n^3$$

3. Now, we return to the main expression:

$$n^3 + n^4 - mn^4 - 2mn^3 - m^2n^3 - n^2(n^2 - m^3n^3 - 2mn) - m(n^5m^2 - mn^3 - n^4)$$
$$= n^3 + n^4 - mn^4 - 2mn^3 - m^2n^3 - n^4 + m^3n^5 + 2mn^3 - n^5m^3 + m^2n^3 + mn^4 = \boxed{n^3}$$

(h)

$$\sqrt{b^3}\left(\sqrt{b^5} - \sqrt{b^3} + \sqrt[3]{b}\right) - \sqrt[6]{b^{11}} - b^3(b-1) = b^{\frac{3}{2}}\left(b^{\frac{5}{2}} - b^{\frac{3}{2}} + b^{\frac{1}{3}}\right) - b^{\frac{11}{6}} - b^3(b-1)$$
$$= b^{\frac{3}{2}}b^{\frac{5}{2}} - b^{\frac{3}{2}}b^{\frac{3}{2}} + b^{\frac{3}{2}}b^{\frac{1}{3}} - b^{\frac{11}{6}} - b^4 + b^3 = b^4 - b^3 + b^{\frac{11}{6}} - b^{\frac{11}{6}} - b^4 + b^3 = \boxed{0}$$

(i)

$$\sqrt{T^3}\left(\sqrt{T^3}\left(2\sqrt[3]{T^5} - 3\sqrt[4]{T^3}\right)\right) + T^3\left(3\sqrt[4]{T^3} - 2\sqrt[3]{T^5}\right)$$
$$= T^{\frac{3}{2}}\left(T^{\frac{3}{2}}\left(2T^{\frac{5}{3}} - 3T^{\frac{3}{4}}\right)\right) + T^3\left(3T^{\frac{3}{4}} - 2T^{\frac{5}{3}}\right) = T^{\frac{3}{2}}T^{\frac{3}{2}}\left(2T^{\frac{5}{3}} - 3T^{\frac{3}{4}}\right) + 3T^3T^{\frac{3}{4}} - 2T^3T^{\frac{5}{3}}$$
$$= 2T^{\frac{3}{2}}T^{\frac{3}{2}}T^{\frac{5}{3}} - 3T^{\frac{3}{2}}T^{\frac{3}{2}}T^{\frac{3}{4}} + 3T^3T^{\frac{3}{4}} - 2T^3T^{\frac{5}{3}} = 2T^{\frac{14}{3}} - 3T^{\frac{15}{4}} + 3T^{\frac{15}{4}} - 2T^{\frac{14}{3}} = \boxed{0}$$

(j)

$$\sqrt[3]{f^5}\left(2\sqrt[3]{f^4} - 3\sqrt[4]{f^3}\right) - 3\sqrt[4]{f^3}\left(\sqrt[4]{f^5} - \sqrt[3]{f^5}\right) - \left(2f^3 - 3\sqrt[4]{f^5}\left(\sqrt[4]{f^3} + 1\right)\right)$$
$$= 2f^{\frac{5}{3}}f^{\frac{4}{3}} - 3f^{\frac{5}{3}}f^{\frac{3}{4}} - 3f^{\frac{3}{4}}f^{\frac{5}{4}} + 3f^{\frac{3}{4}}f^{\frac{5}{3}} - 2f^3 + 3f^{\frac{5}{4}}f^{\frac{3}{4}} + 3f^{\frac{5}{4}}$$
$$= 2f^3 - 3f^{\frac{29}{12}} - 3f^2 + 3f^{\frac{29}{12}} - 2f^3 + 3f^2 + 3f^{\frac{5}{4}} = 3f^{\frac{5}{4}} = \boxed{3\sqrt[4]{f^5}}$$

Exercise 3.11

(a)

$$\frac{x}{2} + \frac{x+y}{x} = \frac{x}{2}\cdot\frac{x}{x} + \frac{x+y}{x}\cdot\frac{2}{2} = \frac{x^2}{2x} + \frac{2(x+y)}{2x} = \boxed{\frac{x^2 + 2x + 2y}{2x}}$$

(b)

$$\frac{3}{y} - \frac{4}{x-y} = \frac{3}{y}\cdot\frac{x-y}{x-y} - \frac{4}{x-y}\cdot\frac{y}{y} = \frac{3(x-y)}{y(x-y)} - \frac{4y}{y(x-y)} = \frac{3x - 3y - 4y}{yx - y^2} = \boxed{\frac{3x - 7y}{yx - y^2}}$$

(c)

$$\frac{c^2}{1+c} + \frac{5}{c^2} = \frac{c^2}{1+c} \cdot \frac{c^2}{c^2} + \frac{5}{c^2} \cdot \frac{1+c}{1+c} = \frac{c^2 c^2}{c^2(1+c)} + \frac{5(1+c)}{c^2(1+c)} = \boxed{\frac{c^4 + 5c + 5}{c^3 + c^2}}$$

(d)

$$\frac{2a^2}{a+b} + \frac{3b}{a^2} = \frac{2a^2}{a+b} \cdot \frac{a^2}{a^2} + \frac{3b}{a^2} \cdot \frac{a+b}{a+b} = \frac{2a^4}{a^2(a+b)} + \frac{3b(a+b)}{a^2(a+b)} = \boxed{\frac{2a^4 + 3ab + 3b^2}{a^3 + a^2 b}}$$

(e)

$$-\frac{4p^2 + p}{p-q} - \frac{3q}{p^2} = -\frac{4p^2 + p}{p-q} \cdot \frac{p^2}{p^2} - \frac{3q}{p^2} \cdot \frac{p-q}{p-q} = -\frac{(4p^2 + p)p^2}{(p-q)p^2} - \frac{3q(p-q)}{p^2(p-q)}$$

$$= \boxed{\frac{-4p^4 - p^3 + 3q^2 - 3qp}{p^3 - p^2 q}}$$

(f)

$$\frac{3A^2 + 2B^3}{A+B} + \frac{A}{B} = \frac{3A^2 + 2B^3}{A+B} \cdot \frac{B}{B} + \frac{A}{B} \cdot \frac{A+B}{A+B} = \frac{B(3A^2 + 2B^3)}{B(A+B)} + \frac{A(A+B)}{B(A+B)}$$

$$= \boxed{\frac{3A^2 B + 2B^4 + A^2 + AB}{AB + B^2}}$$

(g)

$$\frac{3x^2}{x+y} + \frac{4y}{x^2} - \frac{2x}{y^2} = \frac{3x^2}{x+y} \cdot \frac{x^2 y^2}{x^2 y^2} + \frac{4y}{x^2} \cdot \frac{(x+y)y^2}{(x+y)y^2} - \frac{2x}{y^2} \cdot \frac{(x+y)x^2}{(x+y)x^2}$$

$$= \frac{3x^4 y^2}{x^2 y^2 (x+y)} + \frac{4y^3(x+y)}{x^2 y^2 (x+y)} - \frac{2x^3(x+y)}{x^2 y^2 (x+y)} = \boxed{\frac{3x^4 y^2 + 4xy^3 + 4y^4 - 2x^4 - 2x^3 y}{x^3 y^2 + x^2 y^3}}$$

(h)

$$\frac{5m^2}{n} - \frac{7n}{m^2} - \frac{3mn}{m^2 + n^2} = \frac{5m^2}{n} \cdot \frac{m^2(m^2 + n^2)}{m^2(m^2 + n^2)} - \frac{7n}{m^2} \cdot \frac{n(m^2 + n^2)}{n(m^2 + n^2)} - \frac{3mn}{m^2 + n^2} \cdot \frac{m^2 n}{m^2 n}$$

$$= \frac{5m^4(m^2 + n^2)}{m^2 n(m^2 + n^2)} - \frac{7n^2(m^2 + n^2)}{m^2 n(m^2 + n^2)} - \frac{3m^3 n^2}{m^2 n(m^2 + n^2)}$$

$$= \boxed{\frac{5m^6 + 5m^4 n^2 - 7n^2 m^2 - 7n^4 - 3m^3 n^2}{m^4 n + m^2 n^3}}$$

Exercise 3.12

(a)

$$\sqrt{\frac{x^3 + x^2 + x^2 y^2}{x + y^2 + 1}} = \sqrt{\frac{x^2(x + 1 + y^2)}{x + 1 + y^2}} = \sqrt{x^2 \cdot \frac{x + 1 + y^2}{x + 1 + y^2}} = \sqrt{x^2} = \boxed{|x|}$$

Here $|x|$ is the absolute value of x.

(b)

$$\frac{AB^2 + 4B^2}{A^2 B + 4AB} = \frac{B^2(A + 4)}{AB(A + 4)} = \frac{B}{A} \cdot \frac{B}{B} \cdot \frac{A + 4}{A + 4} = \boxed{\frac{B}{A}}$$

(c)

$$\frac{9x^2 y - 18xy^2}{9x^2 - 18xy} = \frac{9xy(x - 2y)}{9x(x - 2y)} = y \cdot \frac{9}{9} \cdot \frac{x}{x} \cdot \frac{x - 2y}{x - 2y} = \boxed{y}$$

(d)

$$\frac{x^4 y - 8x^3 y^2}{yx^4 - 4y^2 x^3} - \frac{x - 4y - 4y}{x - y - 3y} = \frac{x^4 y - 8x^3 y^2}{yx^4 - 4y^2 x^3} - \frac{x - 8y}{x - 4y}$$

$$= \frac{x^3 y(x - 4y)}{x^3 y(x - 8y)} - \frac{x - 8y}{x - 4y} = \frac{x - 8y}{x - 4y} - \frac{x - 8y}{x - 4y} = \boxed{0}$$

(e)

$$\frac{m^3 n^2 - m^2 n^3}{m^4 n^2 - m^2 n^2} \cdot \frac{m^2 - 1}{m - n} = \frac{m^2 n^2(m - n)}{m^2 n^2(m^2 - 1)} \cdot \frac{m^2 - 1}{m - n} = \boxed{1}$$

(f)

$$\frac{P^3 Q^3 - PQ^4}{P^2 Q^3 - PQ^4} - \frac{Q - PQ}{Q - P} = \frac{PQ^3(P^2 - Q)}{PQ^3(P - Q)} - \frac{Q - PQ}{Q - P} = \frac{P^2 - Q}{P - Q} + \frac{Q - PQ}{P - Q}$$

$$= \frac{P^2 - Q + Q - PQ}{P - Q} = \frac{P^2 - PQ}{P - Q} = \frac{P(P - Q)}{P - Q} = \boxed{P}$$

(g)

$$\sqrt{\frac{2a^2 b^2 - 4ab^3 + 2b^4}{2a^2 - 4ab + 2b^2}} = \sqrt{\frac{2b^2(a^2 - 2ab + b^2)}{2(a^2 - 2ab + b^2)}} = \sqrt{b^2} = \boxed{|b|}$$

(h)

$$\sqrt{\frac{(t^2 u)^2 - t^3 u^3 - t^3 u^3 + (tu^2)^2}{t^2 - 2tu + u^2}} = \sqrt{\frac{t^4 u^2 - 2t^3 u^3 + t^2 u^4}{t^2 - 2tu + u^2}}$$

$$= \sqrt{\frac{t^2 u^2(t^2 - 2tu + u^2)}{t^2 - 2tu + u^2}} = \sqrt{(tu)^2} = \boxed{|tu|}$$

(i)

$$\sqrt[4]{\frac{v^2(u-2v)^4 - u^2(u-2v)^4}{v^2 - u^2}} = \sqrt[4]{\frac{(u-2v)^4(v^2 - u^2)}{v^2 - u^2}} = \sqrt[4]{(u-2v)^4} = \boxed{|u-2v|}$$

(j)

$$\frac{wz + z + (wz + z)^3 + (wz + z)^5}{z + z(wz + z)^2 + z(wz + z)^4} = \frac{(wz + z)(1 + (wz + z)^2 + (wz + z)^4)}{z(1 + (wz + z)^2 + (wz + z)^4)}$$
$$= \frac{(wz + z)}{z} = \frac{z(w + 1)}{z} = \boxed{w + 1}$$

Exercise 3.13

(a)

$$\frac{a(a+b)}{a^2 + ba} + \frac{6(x - z - t)}{3x - 3t - 3z} - \frac{mn - mn^2 - mn^3}{mn(1 - n^2 - n)}$$
$$= \frac{a(a+b)}{a(a+b)} + \frac{6(x - z - t)}{3(x - z - t)} + \frac{mn(1 - n - n^2)}{mn(1 - n^2 - n)} = 1 + 2 - 1 = \boxed{2}$$

(b)

$$\frac{y^5(6x^3 - 4y^2)}{x^2(xy^5 + 2y^6)} \cdot \frac{x^3 + 2x^2y}{2y^2 - 3x^3} = -\frac{2y^5(2y^2 - 3x^3)}{x^2y^5(x + 2y)} \cdot \frac{x^2(x + 2y)}{2y^2 - 3x^3}$$
$$= -2 \cdot \frac{y^5}{y^5} \cdot \frac{x^2}{x^2} \cdot \frac{2y^2 - 3x^3}{2y^2 - 3x^3} \cdot \frac{x + 2y}{x + 2y} = \boxed{-2}$$

(c)

$$\frac{f^2(3g(f + g^2f))}{f^3(g^3 + g)} \cdot \frac{f(g^4 - fg^3) - g^4}{g^3(fg - f^2 - g)} = 3 \cdot \frac{f^3g(1 + g^2)}{f^3g(1 + g^2)} \cdot \frac{fg^4 - f^2g^3 - g^4}{fg^4 - f^2g^3 - g^4} = \boxed{3}$$

(d)

$$\frac{xw^2 + xw}{x + w} \cdot \frac{x^2 - x^2w}{xw(w - 1)} \cdot \frac{w + x}{(w + 1)x^2} = \frac{xw(w + 1)}{x + w} \cdot \frac{x^2(1 - w)}{xw(w - 1)} \cdot \frac{w + x}{(w + 1)x^2}$$
$$= -\frac{x^3w(w + 1)(1 - w)(w + x)}{x^3w(w + 1)(1 - w)(w + x)} = \boxed{-1}$$

(e) For clarity, let's denote $A = (x + z)^5$:

$$1 - A\left(\frac{A + y}{yA} - \frac{y}{A} - \frac{1}{y}\right) = 1 - A\left(\frac{A + y}{yA} - \frac{y}{A} \cdot \frac{y}{y} - \frac{1}{y} \cdot \frac{A}{A}\right)$$
$$= 1 - A \cdot \frac{A + y - y^2 - A}{yA} = 1 - \frac{A(y - y^2)}{yA} = 1 - (1 - y)\frac{Ay}{Ay}$$
$$= 1 - 1 + y = \boxed{y}$$

(f)

$$\left(\frac{xy}{z} + \frac{y}{z-1}\right)\frac{z(z-1)}{z-x+xz} = \left(\frac{xy}{z}\cdot\frac{z-1}{z-1} + \frac{y}{z-1}\cdot\frac{z}{z}\right)\frac{z(z-1)}{z-x+xz}$$

$$= \frac{xy(z-1)+yz}{z(z-1)}\cdot\frac{z(z-1)}{z-x+xz} = \frac{xyz-xy+yz}{z-x+xz} = \frac{y(xz-x+z)}{z-x+xz} = \boxed{y}$$

(g)

$$\frac{rt-s(r+s)}{st^2}\left(\frac{r-s-t}{st} - \frac{r+s}{t^2} + \frac{1}{s} + \frac{1}{t}\right)^{-1}$$

$$= \frac{rt-rs-s^2}{st^2}\left(\frac{r-s-t}{st}\cdot\frac{t}{t} - \frac{r+s}{t^2}\cdot\frac{s}{s} + \frac{1}{s}\cdot\frac{t^2}{t^2} + \frac{1}{t}\cdot\frac{st}{st}\right)^{-1}$$

$$= \frac{rt-rs-s^2}{st^2}\left(\frac{(r-s-t)t}{st^2} - \frac{(r+s)s}{st^2} + \frac{t^2}{st^2} + \frac{st}{st^2}\right)^{-1}$$

$$= \frac{rt-rs-s^2}{st^2}\left(\frac{rt-st-t^2-rs-s^2+t^2+st}{st^2}\right)^{-1}$$

$$= \frac{rt-rs-s^2}{st^2}\left(\frac{rt-rs-s^2}{st^2}\right)^{-1} = \frac{rt-rs-s^2}{st^2}\cdot\frac{st^2}{rt-rs-s^2} = \boxed{1}$$

(h)

$$\frac{x(y+z)}{xy+xz} - \frac{z^{-2}}{x^2} + \frac{x^2+y^2}{(xyz)^2} - \frac{y^{-2}}{z^2} = \frac{xy+xz}{xy+xz} - \frac{1}{x^2z^2} + \frac{x^2+y^2}{x^2y^2z^2} - \frac{1}{y^2z^2}$$

$$= 1 - \frac{1}{x^2z^2} + \frac{x^2+y^2}{x^2y^2z^2} - \frac{1}{y^2z^2} = \frac{x^2y^2z^2}{x^2y^2z^2} - \frac{1}{x^2z^2}\cdot\frac{y^2}{y^2} + \frac{x^2+y^2}{x^2y^2z^2} - \frac{1}{y^2z^2}\cdot\frac{x^2}{x^2}$$

$$= \frac{x^2y^2z^2-y^2+x^2+y^2-x^2}{x^2y^2z^2} = \frac{x^2y^2z^2}{x^2y^2z^2} = \boxed{1}$$

(i)

$$\frac{t^3(t+s-2)-s^2(t+3)}{s(s(-t-3)+t^3)+(t-2)t^3} = \frac{t^4+t^3s-2t^3-s^2t-3s^2}{-s^2t-3s^2+st^3+t^4-2t^3}$$

$$= \frac{t^4+t^3s-2t^3-s^2t-3s^2}{t^4+t^3s-2t^3-s^2t-3s^2} = \boxed{1}$$

(j) Let's denote $A = (H+3)^3$:

$$\frac{G^5(T+2A)-GT^3(5A+G)}{G(G^3(2A+T)-T^3)-5AT^3} = \frac{G^5T+2G^5A-5GT^3A-G^2T^3}{2G^4A+G^4T-GT^3-5AT^3}$$

$$= \frac{G(G^4T+2G^4A-5T^3A-GT^3)}{2G^4A+G^4T-GT^3-5AT^3} = \frac{G(G^4T+2G^4A-5T^3A-GT^3)}{G^4T+2G^4A-5T^3A-GT^3} = \boxed{G}$$

(k)

$$\frac{\alpha^2\beta(1-\alpha^2)-\alpha^3\beta^2(1-\beta^2)}{\alpha(\beta^2(\beta^2-1)-\alpha\beta)+\beta}\left(\frac{\alpha^2+\alpha(\beta+\beta^4)}{\beta^4+\alpha+\beta}\right)^{-1}$$

$$=\frac{\alpha^2\beta-\alpha^4\beta-\alpha^3\beta^2+\alpha^3\beta^4}{\alpha\beta^4-\alpha\beta^2-\alpha^2\beta+\beta}\left(\frac{\alpha^2+\alpha\beta+\alpha\beta^4}{\beta^4+\alpha+\beta}\right)^{-1}$$

$$=\frac{\alpha^2\beta-\alpha^4\beta-\alpha^3\beta^2+\alpha^3\beta^4}{\alpha\beta^4-\alpha\beta^2-\alpha^2\beta+\beta}\cdot\frac{\beta^4+\alpha+\beta}{\alpha^2+\alpha\beta+\alpha\beta^4}$$

$$=\frac{\alpha^2\beta(1-\alpha^2-\alpha\beta+\alpha\beta^3)}{\beta(\alpha\beta^3-\alpha\beta-\alpha^2+1)}\cdot\frac{\beta^4+\alpha+\beta}{\alpha(\alpha+\beta+\beta^4)}=$$

$$\frac{\alpha^2\beta(1-\alpha^2-\alpha\beta+\alpha\beta^3)(\beta^4+\alpha+\beta)}{\alpha\beta(1-\alpha^2-\alpha\beta+\alpha\beta^3)(\beta^4+\alpha+\beta)}=\boxed{\alpha}$$

(l) Let's denote $A=20m+m^2$

$$\sqrt[3]{\left(\frac{n^2}{n+1}+\frac{n^3}{n^3+n^2}\right)^2}\cdot\sqrt{\frac{\sqrt[3]{n^4}\left(\sqrt{A^3}+\sqrt[3]{n^5}\sqrt[3]{A^2}\right)}{\sqrt[3]{A^2n^2}\left(\sqrt[6]{A^5}+\sqrt[3]{n^5}\right)}}$$

$$=\left(\frac{n^2+n}{n+1}\right)^{\frac{2}{3}}\left(\frac{n^{\frac{4}{3}}\left(A^{\frac{3}{2}}+n^{\frac{5}{3}}A^{\frac{2}{3}}\right)}{A^{\frac{2}{3}}n^{\frac{2}{3}}\left(A^{\frac{5}{6}}+n^{\frac{5}{3}}\right)}\right)^{\frac{1}{2}}=\left(\frac{n(n+1)}{n+1}\right)^{\frac{2}{3}}\left(\frac{n^{\frac{4}{3}}A^{\frac{2}{3}}\left(A^{\frac{5}{6}}+n^{\frac{5}{3}}\right)}{A^{\frac{2}{3}}n^{\frac{2}{3}}\left(A^{\frac{5}{6}}+n^{\frac{5}{3}}\right)}\right)^{\frac{1}{2}}$$

$$=n^{\frac{2}{3}}\left(n^{\frac{2}{3}}\right)^{\frac{1}{2}}=n^{\frac{2}{3}}n^{\frac{1}{3}}=\boxed{n}$$

Exercise 3.14

1. Simplify the fraction inside A:

$$\frac{\sqrt{(uv)^3}\left(\sqrt{u}+\sqrt{v^3}-uv\left(\sqrt{u^3}+\sqrt{v}\right)\right)}{\left(\sqrt{u}+\sqrt{v^3}\right)\sqrt{(uv)^5}-\left(\sqrt{v}+\sqrt{u^3}\right)\sqrt{(uv)^7}}=$$

$$\frac{\sqrt{(uv)^3}\left(\sqrt{u}+\sqrt{v^3}-v\sqrt{u^5}-u\sqrt{v^3}\right)}{\sqrt{(uv)^5}\left(\sqrt{u}+\sqrt{v^3}-u\sqrt{v^3}-v\sqrt{u^5}\right)}=\frac{1}{uv}$$

2. Simplify the original expression:

$$\sqrt{A^{-1}B} = \sqrt{\left(\frac{1}{uv} + u^{-2} + v^{-2}\right)^{-1}\left(\frac{u+v}{u} + \frac{v^2}{u^2}\right)}$$

$$= \sqrt{\left(\frac{1}{uv} + \frac{1}{u^2} + \frac{1}{v^2}\right)^{-1}\left(\frac{u+v}{u} + \frac{v^2}{u^2}\right)}$$

$$= \sqrt{\left(\frac{1}{uv}\cdot\frac{uv}{uv} + \frac{1}{u^2}\cdot\frac{v^2}{v^2} + \frac{1}{v^2}\cdot\frac{u^2}{u^2}\right)^{-1}\left(\frac{u+v}{u}\cdot\frac{u}{u} + \frac{v^2}{u^2}\right)}$$

$$= \sqrt{\left(\frac{uv + v^2 + u^2}{u^2v^2}\right)^{-1}\cdot\frac{(u+v)u + v^2}{u^2}} = \sqrt{\frac{u^2v^2}{uv + v^2 + u^2}\cdot\frac{u^2 + uv + v^2}{u^2}}$$

$$= \sqrt{v^2} = \boxed{v}$$

Exercise 3.15

1. Simplify the leftmost expression:

$$\left(\frac{x^2}{y} - \frac{x^3 - y^2}{xy}\right)^{-1} = \left(\frac{x^2}{y} - \frac{x^3 - y^2}{xy}\right)^{-1} = \left(\frac{x^2}{y}\cdot\frac{x}{x} - \frac{x^3 - y^2}{xy}\right)^{-1}$$

$$= \left(\frac{x^3 - x^3 + y^2}{xy}\right)^{-1} = \left(\frac{y}{x}\right)^{-1} = \frac{x}{y}$$

2. Work on the expression in the middle:

$$\frac{2x+y}{x+3y^2} - \frac{1}{xy} - \frac{y}{x} = \frac{2x+y}{x+3y^2}\cdot\frac{xy}{xy} - \frac{1}{xy}\cdot\frac{x+3y^2}{x+3y^2} - \frac{y}{x}\cdot\frac{y(x+3y^2)}{y(x+3y^2)}$$

$$= \frac{2x^2y + xy^2 - x - 3y^2 - xy^2 - 3y^4}{xy(x+3y^2)} = \frac{2x^2y - x - 3y^2 - 3y^4}{xy(x+3y^2)}$$

3. Work on the rightmost expression:

$$\frac{(2y)^2 + x - y^2}{2\left(x - \frac{1}{2}y^{-1}\right) - 3yx^{-1}(y^2+1)} = \frac{x + 3y^2}{2x - \frac{1}{y} - \frac{3y^3}{x} - \frac{3y}{x}} = \frac{x + 3y^2}{2x - \frac{1}{y} - \frac{3y^3}{x} - \frac{3y}{x}}$$

$$= \frac{xy(x+3y^2)}{2x^2y - x - 3y^4 - 3y^2}$$

4. Return to the original expression:

$$\frac{x}{y} \cdot \frac{2x^2y - x - 3y^2 - 3y^4}{xy(x + 3y^2)} \cdot \frac{xy(x + 3y^2)}{2x^2y - x - 3y^2 - 3y^4} = \boxed{\frac{x}{y}}$$

5. Evaluate for the given values:

$$\frac{x}{y} = \frac{8}{4} = \boxed{2}$$

Exercise 3.16

$$\left(\frac{ab\left(a\left(b\right)^2\right)^{-2}}{a + b + 2} - b\left(\frac{a}{b}\right)^{-2} - \left(\frac{1}{a}\right)^{-1} + \frac{a\left(b^6 - 1\right) + (b + 2)b^6}{a^2b^{-1}(ab^4 + (b + 2)b^4)} \right)^{2b}$$

$$= \left(\frac{1}{ab^3(a + b + 2)} - \frac{b^3}{a^2} - a + \frac{ab^6 - a + b^7 + 2b^6}{a^2b^3(a + b + 2)} \right)^{2b}$$

$$= \left(\frac{a - ab^6 - b^7 - 2b^6 - a^4b^3 - a^3b^4 - 2a^3b^3 + ab^6 - a + b^7 + 2b^6}{a^2b^3(a + b + 2)} \right)^{2b}$$

$$= \left(\frac{a - a^3b^3(a + b + 2) + ab^6 - a + b^7 + 2b^6}{a^2b^3(a + b + 2)} \right)^{2b}$$

$$= \left(\frac{a - ab^6 - b^7 - 2b^6 - a^4b^3 - a^3b^4 - 2a^3b^3 + ab^6 - a + b^7 + 2b^6}{a^2b^3(a + b + 2)} \right)^{2b}$$

$$= \left(-\frac{a^4b^3 + a^3b^4 + 2a^3b^3}{a^2b^3(a + b + 2)} \right)^{2b} = \left(-\frac{a^3b^3(a + b + 2)}{a^2b^3(a + b + 2)} \right)^{2b} = (-a)^{2b} = \boxed{a^{2b}}$$

Evaluate the expression for the given values:

$$a^{2b} = 2^{2 \cdot 3} = \boxed{64}$$

Exercise 3.17

1. Work with A:

$$\frac{\sqrt{t}}{\sqrt[3]{s}} - \frac{\sqrt{st}+\sqrt{t}}{\sqrt[3]{s^2}\sqrt[3]{t^2}} + \frac{\sqrt{s}\sqrt[3]{\frac{1}{t^2}} - \sqrt{st} + \sqrt[6]{\frac{1}{t}}}{\sqrt[3]{s^2}}$$

$$= \frac{t^{\frac{1}{2}}}{s^{\frac{1}{3}}} \cdot \frac{s^{\frac{1}{3}}t^{\frac{2}{3}}}{s^{\frac{1}{3}}t^{\frac{2}{3}}} - \frac{s^{\frac{1}{2}}t^{\frac{1}{2}}+t^{\frac{1}{2}}}{s^{\frac{2}{3}}t^{\frac{2}{3}}} + \frac{s^{\frac{1}{2}}t^{-\frac{2}{3}} - s^{\frac{1}{2}}t^{\frac{1}{2}} + t^{-\frac{1}{6}}}{s^{\frac{2}{3}}} \cdot \frac{t^{\frac{2}{3}}}{t^{\frac{2}{3}}}$$

$$= \frac{t^{\frac{7}{6}}s^{\frac{1}{3}} - s^{\frac{1}{2}}t^{\frac{1}{2}} - t^{\frac{1}{2}} + s^{\frac{1}{2}} - s^{\frac{1}{2}}t^{\frac{7}{6}} + t^{\frac{1}{2}}}{s^{\frac{2}{3}}t^{\frac{2}{3}}}$$

$$= \frac{s^{\frac{1}{3}}t^{\frac{7}{6}} - s^{\frac{1}{2}}t^{\frac{1}{2}} + s^{\frac{1}{2}} - s^{\frac{1}{2}}t^{\frac{7}{6}}}{s^{\frac{2}{3}}t^{\frac{2}{3}}}$$

2. Work with B, always referring to the structure of A for a hint:

$$\frac{t^{\frac{4}{3}}}{s^{-1}\left(-\sqrt[6]{s} - \sqrt[6]{t^7}\right) + \left(t^{\frac{2}{3}}+1\right)s^{-\frac{5}{6}}\sqrt{t}} = \frac{t^{\frac{4}{3}}}{-s^{-\frac{5}{6}} - s^{-1}t^{\frac{7}{6}} + s^{-\frac{5}{6}}t^{\frac{7}{6}} + s^{-\frac{5}{6}}t^{\frac{1}{2}}}$$

$$= \frac{t^{\frac{4}{3}}}{-s^{-\frac{4}{3}}\left(s^{\frac{1}{2}} + s^{\frac{1}{3}}t^{\frac{7}{6}} - s^{\frac{1}{2}}t^{\frac{7}{6}} - s^{\frac{1}{2}}t^{\frac{1}{2}}\right)} = -\frac{s^{\frac{4}{3}}t^{\frac{4}{3}}}{s^{\frac{1}{3}}t^{\frac{7}{6}} - s^{\frac{1}{2}}t^{\frac{1}{2}} + s^{\frac{1}{2}} - s^{\frac{1}{2}}t^{\frac{7}{6}}}$$

3. Simplify the original expression:

$$(AB)^3 = -\left(\frac{s^{\frac{1}{3}}t^{\frac{7}{6}} - s^{\frac{1}{2}}t^{\frac{1}{2}} + s^{\frac{1}{2}} - s^{\frac{1}{2}}t^{\frac{7}{6}}}{s^{\frac{2}{3}}t^{\frac{2}{3}}} \cdot \frac{s^{\frac{4}{3}}t^{\frac{4}{3}}}{s^{\frac{1}{3}}t^{\frac{7}{6}} - s^{\frac{1}{2}}t^{\frac{1}{2}} + s^{\frac{1}{2}} - s^{\frac{1}{2}}t^{\frac{7}{6}}}\right)^3$$

$$= -\left(s^{\frac{2}{3}}t^{\frac{2}{3}}\right)^3 = \boxed{-s^2t^2}$$

4. Evaluate for the given values:

$$-s^2t^2 = -2^2 \cdot 3^2 = \boxed{-36}$$

Exercise 3.18

$$\sqrt{\frac{x + xy + x + 2xy + x + 3xy + \cdots + x + nxy}{nx^{-1} + y(x^{-1}n + x^{-1}(n-1) + x^{-1}(n-2) + \cdots + x^{-1})}}$$

$$= \sqrt{\frac{\underbrace{x + x + x + \cdots + x}_{n \; times} + xy + 2xy + 3xy + \cdots + nxy}{x^{-1}(n + y(1 + 2 + 3 + \cdots + n))}}$$

$$= \sqrt{\frac{x(n + y + 2y + 3y + \cdots + ny)}{x^{-1}(n + y(1 + 2 + 3 + \cdots + n))}}$$

$$= \sqrt{\frac{x^2(n + y(1 + 2 + 3 + \cdots + n))}{n + y(1 + 2 + 3 + \cdots + n)}} = \sqrt{x^2} = \boxed{|x|}$$

Exercise 3.19

(a) 1. Simplify the first term:

$$x(5 + x(4 + x(3 + x(2 + x(1 + x))))) = 5x + x^2(4 + x(3 + x(2 + x(1 + x))))$$
$$= 5x + 4x^2 + x^3(3 + x(2 + x(1 + x))) = 5x + 4x^2 + 3x^3 + x^4(2 + x(1 + x))$$
$$= 5x + 4x^2 + 3x^3 + 2x^4 + x^5(1 + x) = 5x + 4x^2 + 3x^3 + 2x^4 + x^5 + x^6$$

2. Simplify the whole expression:

$$(5x + 4x^2 + 3x^3 + 2x^4 + x^5) + x^6 - (5x + 4x^2 + 3x^3 + 2x^4 + x^5) = \boxed{x^6}$$

(b) The generalized expression will have the following form (we can see the pattern from solving part (a)):

$$(nx + (n-1)x^2 + (n-2)x^3 + \cdots + x^n) + x^{n+1} -$$
$$- (nx + (n-1)x^2 + (n-2)x^3 + \cdots + x^n) = \boxed{x^{n+1}}$$

Exercise 3.20

(a)

$$A + A^* = (1 + 2 + 3 + \cdots + n) + (n + (n-1) + (n-2) + \cdots + 1)$$
$$= (1 + n) + (2 + n - 1) + (3 + n - 2) + \cdots (n + 1)$$
$$= \underbrace{(1 + n) + (1 + n) + \cdots + (1 + n)}_{n \; times}$$

$$(1 + n)(\underbrace{1 + 1 + \cdots + 1}_{n \; times}) = \boxed{n(1 + n)}$$

(b) Since $A = A^*$, we have:

$$2A = A + A^* = n(1 + n)$$

From here:

$$A = \boxed{\frac{n(n + 1)}{2}}$$

(c)

$$\frac{K^2 + 2K^2 + 3K^2 + \cdots + nK^2}{Kn + K} + \frac{Kn}{2} = \frac{K^2(1 + 2 + 3 + \cdots + n)}{K(n + 1)} + \frac{Kn}{2}$$

$$= \frac{K^2 n(n + 1)}{2K(n + 1)} + \frac{Kn}{2} = \frac{Kn}{2} + \frac{Kn}{2} = Kn$$

Evaluate for the given values:

$$Kn = 2 \cdot 1000 = \boxed{2000}$$

Solutions for Chapter 4 Exercises

Exercise 4.1

(a)
$$(3 - x^2)(1 - x^2) = 3 - 3x^2 - x^2 + x^4 = \boxed{x^4 - 4x^2 + 3}$$

(b)
$$(4 - y^2)(2 + y^2) = 8 + 4y^2 - 2y^2 - y^4 = \boxed{-y^4 + 2y^2 + 8}$$

(c)
$$(5\sqrt{z} - 2)(2\sqrt{z} + 3) = 10z + 15\sqrt{z} - 4\sqrt{z} - 6 = \boxed{10z + 11\sqrt{z} - 6}$$

(d)
$$(\alpha - 3\alpha^4)(\alpha + \alpha^4) = \alpha^2 + \alpha^5 - 3\alpha^5 - 3\alpha^8 = \boxed{-3\alpha^8 - 2\alpha^5 + \alpha^2}$$

(e)
$$\left(2 - \beta^{\frac{2}{3}}\right)\left(\beta^{\frac{4}{3}} + 4\right) = 2\beta^{\frac{4}{3}} + 8 - \beta^2 - 4\beta^{\frac{2}{3}} = \boxed{2\beta^{\frac{4}{3}} - 4\beta^{\frac{2}{3}} - \beta^2 + 8}$$

(f)
$$\left(3\sqrt[3]{\gamma^5} - \sqrt[3]{\gamma}\right)\left(-2\sqrt[3]{\gamma^4} + 1\right) = \left(3\gamma^{\frac{5}{3}} - \gamma^{\frac{1}{3}}\right)\left(-2\gamma^{\frac{4}{3}} + 1\right)$$
$$= -6\gamma^3 + 3\gamma^{\frac{5}{3}} + 2\gamma^{\frac{5}{3}} - \gamma^{\frac{1}{3}} = -6\gamma^3 + 5\gamma^{\frac{5}{3}} - \gamma^{\frac{1}{3}} = \boxed{-6\gamma^3 + 5\sqrt[3]{\gamma^5} - \sqrt[3]{\gamma}}$$

(g)
$$(2 - \rho^3)(3\rho^2 + \rho + 1) = (2 - \rho^3)(3\rho^2) + (2 - \rho^3)(\rho) + (2 - \rho^3)$$
$$= \boxed{-3\rho^5 - \rho^4 - \rho^3 + 6\rho^2 + 2\rho + 2}$$

(h)
$$(1 - Y^2)(2Y + Y^3 - 1) = (1 - Y^2)(2Y) + (1 - Y^2)(Y^3) + (1 - Y^2)(-1)$$
$$= \boxed{-Y^5 - Y^3 + Y^2 + 2Y - 1}$$

(i)

$$(4k - k^2 - 2)(3k^2 + k - 1)$$
$$= (4k - k^2 - 2)(3k^2) + (4k - k^2 - 2)(k) + (4k - k^2 - 2)(-1)$$
$$= \boxed{-3k^4 + 11k^3 - k^2 - 6k + 2}$$

(j)

$$(h - 5h^2 + h^3)(h^2 - h + 1)$$
$$= (h - 5h^2 + h^3)(h^2) + (h - 5h^2 + h^3)(-h) + (h - 5h^2 + h^3)(1)$$
$$= \boxed{h^5 - 6h^4 + 7h^3 - 6h^2 + h}$$

(k)

$$(-7X + 3)\left(2X^2 - X^{-1} - X^3\right)$$
$$= (-7X + 3)(2X^2) + (-7X + 3)(-X^{-1}) + (-7X + 3)(-X^3)$$
$$= \boxed{7X^4 - 17X^3 + 6X^2 - 3X^{-1} + 7}$$

(l)

$$\left(-\frac{7}{D^2} - \frac{2}{D^3} + D\right)(3D^2 - D + 1)$$
$$= \left(-\frac{7}{D^2} - \frac{2}{D^3} + D\right)(3D^2) + \left(-\frac{7}{D^2} - \frac{2}{D^3} + D\right)(-D) + \left(-\frac{7}{D^2} - \frac{2}{D^3} + D\right)(1)$$
$$= \boxed{3D^3 - D^2 + D + D^{-1} - 5D^{-2} - 2D^{-3} - 21}$$

Exercise 4.2

(a)
$$(5 + z)^2 - (1 + z)^2 = 25 + 10z + z^2 - (1 + 2z + z^2) = \boxed{8z + 24}$$

(b)
$$(6 + a)^2 - (2 - a)^2 = 36 + 12a + a^2 - (4 - 4a + a^2) = \boxed{16a + 32}$$

(c)
$$(7\sqrt{x} + 2)^2 + (3\sqrt{x} - 4)^2 = 49x + 28\sqrt{x} + 4 + 9x - 24\sqrt{x} + 16 = \boxed{58x + 4\sqrt{x} + 20}$$

(d)

$$(\alpha + 2\alpha^2)^2 - (\alpha - 4\alpha^2)^2 = \alpha^2 + 4\alpha^3 + 4\alpha^4 - (\alpha^2 - 8\alpha^3 + 16\alpha^4) = \boxed{12\alpha^3 - 12\alpha^4}$$

(e)

$$(\sqrt{V} + \sqrt{K})^2 - (2\sqrt{V} - 2\sqrt{K})^2 = V + 2\sqrt{VK} + K - (4V - 8\sqrt{VK} + 4K)$$
$$= \boxed{-3V + 10\sqrt{KV} - 3K}$$

(f)

$$(2\theta + \rho^2)^2 - (3\theta - 2\rho^2)^2 = 4\theta^2 + 4\theta\rho^2 + \rho^4 - (9\theta^2 - 12\theta\rho^2 + 4\rho^4) = \boxed{-5\theta^2 + 16\theta\rho^2 - 3\rho^4}$$

(g)

$$(1 + s)^2 + (1 - s)^2 - (2 + s)^2 = 1 + 2s + s^2 + 1 - 2s + s^2 - (4 + 4s + s^2) = \boxed{s^2 - 4s - 2}$$

(h)

$$(r^2 + r)^2 - (r^2 - r)^2 - (r^2 - 1)^2 = r^4 + 2r^3 + r^2 - (r^4 - 2r^3 + r^2) - (r^4 - 2r^2 + 1)$$
$$= \boxed{-r^4 + 4r^3 + 2r^2 - 1}$$

(i)

$$(1 + L)^4 = (1 + L)^2(1 + L)^2 = (1 + 2L + L^2)(1 + 2L + L^2)$$
$$= (1 + 2L + L^2) + (1 + 2L + L^2)(2L) + (1 + 2L + L^2)(L^2)$$
$$= \boxed{L^4 + 4L^3 + 6L^2 + 4L + 1}$$

(j)

$$(2 + m)^4 - (2 - m)^4 = (2 + m)^2(2 + m)^2 - (2 - m)^2(2 - m)^2$$
$$= (16 + 32m + 24m^2 + 8m^3 + m^4) - (16 - 32m + 24m^2 - 8m^3 + m^4)$$
$$= \boxed{16m^3 + 64m}$$

(k)

$$(1 - u)^4 - (u^2 - 2u)^2 - 2(u^2 - 2u) = (1 - u)^2(1 - u)^2 - (u^2 - 2u)^2 - 2(u^2 - 2u)$$
$$= u^4 - 4u^3 + 6u^2 - 4u + 1 - (u^4 - 4u^3 + 4u^2) - (2u^2 - 4u) = \boxed{1}$$

(l)

$$(1-F)^6 = (1-F)^2(1-F)^2(1-F)^2 = (1-2F+F^2)(1-2F+F^2)(1-2F+F^2)$$
$$= \left[(1-2F+F^2) - (1-2F+F^2)(2F) + (1-2F+F^2)(F^2)\right](1-2F+F^2)$$
$$= (F^4 - 4F^3 + 6F^2 - 4F + 1)(1-2F+F^2)$$
$$= (F^4-4F^3+6F^2-4F+1) - (F^4-4F^3+6F^2-4F+1)(2F) + (F^4-4F^3+6F^2-4F+1)(F^2)$$
$$= \boxed{F^6 - 6F^5 + 15F^4 - 20F^3 + 15F^2 - 6F + 1}$$

Exercise 4.3

(a)
$$(x-2)(x+2) - (x+3)^2 = x^2 - 4 - (x^2+6x+9) = \boxed{-6x-13}$$

(b)
$$\left(\sqrt{q} - \sqrt{2p}\right)^2 - (\sqrt{q} - \sqrt{p})(\sqrt{q} + \sqrt{p}) = q - 2\sqrt{2}\sqrt{pq} + 2p - (q-p) = \boxed{3p - 2\sqrt{2pq}}$$

(c)
$$(2-f^2)(2+f^2) - (3-f^2)(f^2+3) = 4 - f^4 - (9-f^4) = \boxed{-5}$$

(d)
$$\left(2z^2 - \sqrt{z}\right)\left(\sqrt{z} + 2z^2\right) - \left(z^2 + \sqrt{z}\right)^2 = 4z^4 - z - (z^4 + 2z^2\sqrt{z} + z)$$
$$= \boxed{-2z^{5/2} + 3z^4 - 2z}$$

(e)
$$\left[\left(\sqrt{G} - \sqrt{2H}\right)\left(\sqrt{G} + \sqrt{2H}\right)\right]^2 = (G - 2H)^2 = \boxed{G^2 - 4GH + 4H^2}$$

(f)
$$[(\beta^2 + \beta^3)(\beta^2 - \beta^3)]^2 = (\beta^4 - \beta^6)^2 = \boxed{\beta^{12} - 2\beta^{10} + \beta^8}$$

(g)
$$(1-h)(1+h)(1+h^2) = (1-h^2)(1+h^2) = \boxed{-h^4 + 1}$$

(h)
$$\left(\sqrt{k} - \sqrt{2}\right)\left(\sqrt{k} + \sqrt{2}\right)(k+2) = (k-2)(k+2) = \boxed{k^2 - 4}$$

(i)
$$(2-z)(2+z)(4+z^2) - (4-z^2)^2 = 4 - z^2 - (16 - 2z^2 + z^4) = \boxed{-2z^4 + 8z^2}$$

(j)

$$(1-\theta)^4(1+\theta)^4 - (\theta^2+1)^4$$
$$= (1-\theta^2)^4 - (1+\theta^2)^4 = \left[(1-\theta^2)^2 - (1+\theta^2)^2\right]\left[(1-\theta^2)^2 + (1+\theta^2)^2\right]$$
$$= -4\theta^2(2\theta^4+2) = \boxed{-8\theta^6 - 8\theta^2}$$

(k)

$$(3-l-l^2)(3+l+l^2) + (l-l^2)^2$$
$$= (3-(l+l^2))(3+(l+l^2)) + (l-l^2)^2 = 9 - (l+l^2)^2 + (l-l^2)^2 = \boxed{-4l^3 + 9}$$

(l)

$$\left(\sqrt{d^3} - \sqrt{d} + 2\right)\left(\sqrt{d^3} - 2 + \sqrt{d}\right) = \left(\sqrt{d^3} - (\sqrt{d}-2)\right)\left(\sqrt{d^3} + (\sqrt{d}-2)\right)$$
$$= d^3 - (\sqrt{d}-2)^2 = \boxed{d^3 - d + 4\sqrt{d} - 4}$$

Exercise 4.4

(a)

$$(d-2)^3 - (d+3)^3 = d^3 - 6d^2 + 12d - 8 - (d^3 + 9d^2 + 27d + 27) = \boxed{-15d^2 - 15d - 35}$$

(b)

$$\left(\sqrt[3]{t} - \sqrt[3]{s}\right)^3 - \left(\sqrt[3]{t} + \sqrt[3]{s}\right)^3$$
$$= 3s^{2/3}t^{1/3} - 3s^{1/3}t^{2/3} - s + t - (3s^{2/3}t^{1/3} + 3s^{1/3}t^{2/3} + s + t)$$
$$= \boxed{-6\sqrt[3]{st^2} - 2s}$$

(c)

$$-(1+x)^3 - (x-1)^3 - (2-x)^3$$
$$= -(x^3 + 3x^2 + 3x + 1) - (x^3 - 3x^2 + 3x - 1) - (-x^3 + 6x^2 - 12x + 8)$$
$$= \boxed{-x^3 - 6x^2 + 6x - 8}$$

(d)

$$(1-c)^3 - (2-c)^3 - (3+c)^3$$
$$= -c^3 + 3c^2 - 3c + 1 - (-c^3 + 6c^2 - 12c + 8) - (c^3 + 9c^2 + 27c + 27)$$
$$= \boxed{-c^3 - 12c^2 - 18c - 34}$$

(e)

$$(3 + z)(9 - 3z + z^2) = (3 + z)(3^2 - 3z + z^2) = \boxed{z^3 + 27}$$

(f)

$$(2 - \sqrt{n})(4 + 2\sqrt{n} + n) = (2 - \sqrt{n})(2^2 + 2\sqrt{n} + (\sqrt{n})^2) = \boxed{-\sqrt{n^3} + 8}$$

(g)

$$(\sqrt{k} + 2f)(4f^2 + k - 2f\sqrt{k}) - \sqrt{k^3} + f^3$$
$$= (\sqrt{k} + 2f)((\sqrt{k})^2 - 2f\sqrt{k} + (2f)^2) - \sqrt{k^3} + f^3$$
$$= \sqrt{k^3} + 8f^3 - \sqrt{k^3} + f^3 = \boxed{9f^3}$$

(h)

$$(2 - v)^2(v^2 + 2v + 4)^2 - 64 - v^6 = [(2 - v)(v^2 + 2v + 2^2)]^2 - 64 - v^6$$
$$= [8 - v^3]^2 - 64 - v^6 = v^6 - 16v^3 + 64 - 64 - v^6 = \boxed{-16v^3}$$

(i)

$$(\gamma^3 - \phi^3)((\gamma + \phi)^2 - \gamma\phi)^{-1} = \frac{\gamma^3 - \phi^3}{(\gamma + \phi)^2 - \gamma\phi} = \frac{(\gamma - \phi)(\phi^2 + \gamma\phi + \gamma^2)}{\phi^2 + \gamma\phi + \gamma^2} = \boxed{\gamma - \phi}$$

(j)

$$(F^3 + 8)^3(F^2 + 4 - 2F)^{-3} - (2 - F)^3 = \left(\frac{F^3 + 2^3}{F^2 - 2F + 2^2}\right)^3 - (2 - F)^3$$
$$= (2 + F)^3 - (2 - F)^3 = F^3 + 6F^2 + 12F + 8 - (-F^3 + 6F^2 - 12F + 8)$$
$$= \boxed{2F^3 + 24F}$$

(k)

$$(\sqrt[4]{u} + \sqrt[4]{v})(u + \sqrt{uv} + v)(\sqrt[4]{u} - \sqrt[4]{v})$$
$$= (\sqrt[4]{u} + \sqrt[4]{v})(\sqrt[4]{u} - \sqrt[4]{v})(u + \sqrt{uv} + v)$$
$$= (\sqrt{u} - \sqrt{v})((\sqrt{u})^2 + \sqrt{u}\sqrt{v} + (\sqrt{v})^2) = \boxed{\sqrt{u^3} - \sqrt{v^3}}$$

(l)

$$(l + t)((l + t)^3 - 3lt(l + t))^{-1}(l^2 + t^2 - lt)$$
$$= \frac{(l + t)(l^2 - lt + t^2)}{(l + t)^3 - 3lt(l + t)} = \frac{l^3 + t^3}{l^3 + 3l^2t + 3lt^2 + t^3 - 3l^2t - 3lt^2} = \frac{l^3 + t^3}{l^3 + t^3} = \boxed{1}$$

Exercise 4.5

(a) It must be that:

$$h^2 + 3h + 2 = ac + ad + bc + bd = (a+b)(c+d)$$

From the first term we see that we can take $a = h$ and $c = h$. With this in mind:

$$h^2 + 3h + 2 = h^2 + hd + bh + bd$$

That is:

$$3h + 2 = hd + bh + bd$$

$$2h + h + 2 = dh + bh + bd$$

We now see that $d = 2, b = 1$. Finally:

$$h^2 + 3h + 2 = \boxed{(h+1)(h+2)}$$

Note another solution, using the distributive property:

$$h^2 + 3h + 2 = h^2 + h + 2h + 2 = h(h+1) + 2(h+1) = (h+1)(h+2)$$

(b) It must be that:

$$x^2 + 2x - 3 = ac + ad + bc + bd = (a+b)(c+d)$$

From the first term we see that we can take $a = x$ and $c = x$. With this in mind:

$$x^2 + 2x - 3 = x^2 + xd + bx + bd$$

That is:

$$2x - 3 = xd + bx + bd$$

$$3x - x - 3 = dx + bx + bd$$

We now see that $d = 3, b = -1$. Finally:

$$x^2 + 2x - 3 = \boxed{(x-1)(x+3)}$$

Note another solution, using the distributive property:

$$x^2 + 2x - 3 = x^2 + 3x - x - 3 = x(x+3) - (x+3) = (x+3)(x-1)$$

(c) It must be that:

$$t^2 - 5t + 6 = ac + ad + bc + bd = (a+b)(c+d)$$

From the first term we see that we can take $a = t$ and $c = t$. With this in mind:

$$t^2 - 5t + 6 = t^2 + td + bt + bd$$

That is:

$$-5t + 6 = td + bt + bd$$

$$-2t - 3t + 6 = dt + bt + bd$$

We now see that $d = -2, b = -3$. Finally:

$$t^2 - 5t + 6 = \boxed{(t-3)(t-2)}$$

Note another solution, using the distributive property:

$$t^2 - 5t + 6 = t^2 - 2t - 3t + 6 = t(t-2) - 3(t-2) = (t-3)(t-2)$$

In our following solutions, we will not mention such approach anymore.

(d) It must be that:

$$d - 7\sqrt{d} + 10 = ac + aD + bc + bD = (a+b)(c+D)$$

From the first term we see that we can take $a = \sqrt{d}$ and $c = \sqrt{d}$. With this in mind:

$$d - 7\sqrt{d} + 10 = d + \sqrt{d}D + b\sqrt{d} + bD$$

That is:

$$-7\sqrt{d} + 10 = \sqrt{d}D + b\sqrt{d} + bD$$

$$-5\sqrt{d} - 2\sqrt{d} + 10 = D\sqrt{d} + b\sqrt{d} + bD$$

We now see that $D = -5, b = -2$. Finally:

$$d - 7\sqrt{d} + 10 = \boxed{(\sqrt{d} - 2)(\sqrt{d} - 5)}$$

(e) It must be that:

$$fg + 2f - 2g^2 - 4g = ac + ad + bc + bd = (a+b)(c+d)$$

From the first term we see that we can take $a = f$ and $c = g$. With this in mind:

$$fg + 2f - 2g^2 - 4g = fg + fd + bg + bd$$

That is:

$$2f - 2g^2 - 4g = fd + bg + bd$$

We now see that $d = 2, b = -2g$. Finally:

$$fg + 2f - 2g^2 - 4g = \boxed{(f - 2g)(g + 2)}$$

(f) It must be that:

$$2rs + 2rt + s^2 + st = ac + ad + bc + bd = (a + b)(c + d)$$

From the first term we see that we can take $a = 2r$ and $c = s$. With this in mind:

$$2rs + 2rt + s^2 + st = 2rs + 2rd + bs + bd$$

That is:

$$2rt + s^2 + st = 2rd + bs + bd$$

We now see that $d = t, b = s$. Finally:

$$2rs + 2rt + s^2 + st = \boxed{(2r + s)(s + t)}$$

(g) It must be that:

$$-g + \sqrt{3g} - \sqrt{g} + \sqrt{3} = ac + ad + bc + bd = (a + b)(c + d)$$

From the first term we see that $a = -1$ and $c = g$. With this in mind:

$$-g + \sqrt{3g} - \sqrt{g} + \sqrt{3} = -g - d + bg + bd$$

That is:

$$\sqrt{3g} - \sqrt{g} + \sqrt{3} = -d + bg + bd$$

We now see that $d = -\sqrt{3g}, b = -\frac{1}{\sqrt{g}}$. This gives:

$$-g + \sqrt{3g} - \sqrt{g} + \sqrt{3} = \left(-1 - \frac{1}{\sqrt{g}}\right)(g - \sqrt{3g})$$

This expression already satisfies the requirement. However, we can make a modification to make it a bit simpler:

$$\left(-1 - \frac{1}{\sqrt{g}}\right)(g - \sqrt{3g}) = \left(-1 - \frac{1}{\sqrt{g}}\right)(-\sqrt{g})(\sqrt{3} - \sqrt{g}) = \boxed{(\sqrt{g} + 1)(\sqrt{3} - \sqrt{g})}$$

Note that we could take $\sqrt{3g}$ as the first term at the start of the process (and could obtain that $a = \sqrt{3}$ and $c = \sqrt{g}$), which would simplify the solution. With experience, students develop a "feeling" for the best course of action.

(h) It must be that:

$$-2x^5 - 6x^3 + 4x^2 + 12 = ac + ad + bc + bd = (a + b)(c + d)$$

From the first term we see that we can take $a = -2x^3$ and $c = x^2$. With this in mind:

$$-2x^5 - 6x^3 + 4x^2 + 12 = -2x^5 - 2x^3d + bx^2 + bd$$

That is:

$$-6x^3 + 4x^2 + 12 = -2x^3d + bx^2 + bd$$

We now see that $d = 3, b = 4$. Finally:

$$-2x^5 - 6x^3 + 4x^2 + 12 = \boxed{(-2x^3 + 4)(x^2 + 3)}$$

(i) It must be that:

$$h^{57} - 2h^{34} - h^{23} + 2 = ac + ad + bc + bd = (a + b)(c + d)$$

From the first term we see that we can take $a = h^{34}$ and $c = h^{23}$. With this in mind:

$$h^{57} - 2h^{34} - h^{23} + 2 = h^{57} + h^{34}d + bh^{23} + bd$$

That is:

$$-2h^{34} - h^{23} + 2 = h^{34}d + bh^{23} + bd$$

We now see that $d = -2, b = -1$. Finally:

$$h^{57} - 2h^{34} - h^{23} + 2 = \boxed{(h^{34} - 1)(h^{23} - 2)}$$

(j) It must be that:

$$-R^{50} + R^{35} - R^{30} + R^{15} = ac + ad + bc + bd = (a + b)(c + d)$$

From the first term we see that we can take $a = -R^{15}$ and $c = R^{35}$. With this in mind:

$$-R^{50} + R^{35} - R^{30} + R^{15} = -R^{50} - R^{15}d + bR^{35} + bd$$

That is:

$$R^{35} - R^{30} + R^{15} = -R^{15}d + bR^{35} + bd$$

We now see that $d = -R^{20}, b = -R^{-5}$. Finally:

$$-R^{50} + R^{35} - R^{30} + R^{15} = (-R^{15} - R^{-5})(R^{35} - R^{20})$$

We can simplify the expression we obtained:

$$(-R^{15} - R^{-5})(R^{35} - R^{20}) = (-R^{15} - R^{-5})(-R^5)(R^{15} - R^{30}) = \boxed{(R^{20} + 1)(R^{15} - R^{30})}$$

(k) It must be that:

$$D^{7/3} - D^{5/3}Q^{1/3} + D^{2/3}Q^{1/3} - Q^{2/3} = ac + ad + bc + bd = (a + b)(c + d)$$

From the first term we see that we can take $a = D^{5/3}$ and $c = D^{2/3}$. With this in mind:

$$D^{7/3} - D^{5/3}Q^{1/3} + D^{2/3}Q^{1/3} - Q^{2/3} = D^{7/3} + D^{5/3}d + bD^{2/3} + bd$$

That is:

$$-D^{5/3}Q^{1/3} + D^{2/3}Q^{1/3} - Q^{2/3} = D^{5/3}d + bD^{2/3} + bd$$

We now see that $d = -Q^{1/3}, b = Q^{1/3}$. We arrive at the necessary form:

$$D^{7/3} - D^{5/3}Q^{1/3} + D^{2/3}Q^{1/3} - Q^{2/3} = \boxed{\left(\sqrt[3]{D^5} + \sqrt[3]{Q}\right)\left(\sqrt[3]{D^2} - \sqrt[3]{Q}\right)}$$

(l) It must be that (we have rearranged the order, because it looks reasonable to try $6 = 2 \times 3$):

$$6 + 3v^{7/4} + v^{5/2} + 2v^{3/4} = ac + ad + bc + bd = (a + b)(c + d)$$

From the first term we see that we can take $a = 2$ and $c = 3$. With this in mind:

$$6 + 3v^{7/4} + v^{5/2} + 2v^{3/4} = 6 + 2d + b3 + bd$$

That is:

$$3v^{7/4} + v^{5/2} + 2v^{3/4} = 2d + b3 + bd$$

Looking at the structure of the left and the right side, we will rearrange the order of the left side:

$$2v^{3/4} + 3v^{7/4} + v^{5/2} = 2d + 3b + bd$$

We now see that $d = v^{3/4}, b = v^{7/4}$. We arrive at the necessary form:

$$6 + 3v^{7/4} + v^{5/2} + 2v^{3/4} = (2 + v^{7/4})(3 + v^{3/4}) = \boxed{\left(2 + \sqrt[4]{v^7}\right)\left(3 + \sqrt[4]{v^3}\right)}$$

Exercise 4.6

(a) We have three terms in the expression. This gives us a clue that we are dealing with the square of a sum.

Thus, it must be that:

$$A^2 - 4A + 4 = a^2 + 2ab + b^2 = (a+b)^2$$

$$A^2 + 2 \cdot A \cdot (-2) + (-2)^2 = a^2 + 2ab + b^2 = (a+b)^2$$

This shows that a possible choice is $a = A$ and $b = -2$. Finally:

$$A^2 - 4A + 4 = \boxed{(A-2)^2}$$

Another possible choice is $a = -A$ and $b = 2$. That is

$$A^2 - 4A + 4 = (2-A)^2$$

But of course: $(2-A)^2 = (A-2)^2$.

(b) We have four terms in the expression. This gives us a clue that we are dealing with the cube of a sum.

Thus, it must be that:

$$9\mu^2 + 27\mu + \mu^3 + 27 = a^3 + 3a^2b + 3ab^2 + b^3 = (a+b)^3$$

$$\mu^3 + 3 \cdot \mu^2 \cdot 3 + 3 \cdot \mu \cdot 3^2 + 3^3 = a^3 + 3a^2b + 3ab^2 + b^3$$

This shows that $a = \mu$ and $b = 3$. Finally:

$$9\mu^2 + 27\mu + \mu^3 + 27 = \boxed{(\mu+3)^3}$$

(c) We have three terms in the expression. This gives us a clue that we are dealing with the square of a sum.

Thus, it must be that:

$$2\sqrt{zt} + z + t = a^2 + 2ab + b^2 = (a+b)^2$$

$$(\sqrt{z})^2 + 2 \cdot \sqrt{z} \cdot \sqrt{t} + (\sqrt{t})^2 = a^2 + 2ab + b^2$$

This shows that a possible choice is $a = \sqrt{z}$ and $b = \sqrt{t}$. Finally:

$$2\sqrt{zt} + z + t = \boxed{(\sqrt{z} + \sqrt{t})^2}$$

Another possible choice is $a = -\sqrt{z}$ and $b = -\sqrt{t}$. That is

$$2\sqrt{zt} + z + t = (-\sqrt{z} - \sqrt{t})^2$$

(d) We have four terms in the expression. This gives us a clue that we are dealing with the cube of a sum.

Thus, it must be that:

$$-3d^{2/3}s^{2/3} + 3d^{1/3}s^{4/3} + d - s^2 = a^3 + 3a^2b + 3ab^2 + b^3 = (a + b)^3$$

$$(d^{1/3})^3 + 3 \cdot (d^{1/3})^2 \cdot (-s^{2/3}) + 3 \cdot d^{1/3} \cdot (-s^{2/3})^2 + (-s^{2/3})^3 = a^3 + 3a^2b + 3ab^2 + b^3$$

This shows that $a = d^{1/3}$ and $b = -s^{2/3}$. Finally:

$$-3d^{2/3}s^{2/3} + 3d^{1/3}s^{4/3} + d - s^2 = \boxed{\left(\sqrt[3]{d} - \sqrt[3]{s^2}\right)^3}$$

(e) We have four terms in the expression. This gives us a clue that we are dealing with the cube of a sum.

Thus, it must be that:

$$27f^6 + 54f^4 + 36f^2 + 8 = a^3 + 3a^2b + 3ab^2 + b^3 = (a + b)^3$$

$$(3f^2)^3 + 3 \cdot (3f^2)^2 \cdot 2 + 3 \cdot (3f^2) \cdot 2^2 + 2^3 = a^3 + 3a^2b + 3ab^2 + b^3$$

This shows that $a = 3f^2$ and $b = 2$. Finally:

$$27f^6 + 54f^4 + 36f^2 + 8 = \boxed{\left(3f^2 + 2\right)^3}$$

(f) We have three terms in the expression. This gives us a clue that we are dealing with the square of a sum.

Thus, it must be that:

$$S^{40} + 6S^{30} + 9S^{20} = a^2 + 2ab + b^2 = (a + b)^2$$

$$(S^{20})^2 + 2 \cdot S^{20} \cdot 3S^{10} + (3S^{10})^2 = a^2 + 2ab + b^2$$

This shows that a possible choice is $a = S^{20}$ and $b = 3S^{10}$. Finally:

$$S^{40} + 6S^{30} + 9S^{20} = \boxed{\left(S^{20} + 3S^{10}\right)^2}$$

Another possible choice is $a = -S^{20}$ and $b = -3S^{10}$. That is

$$S^{40} + 6S^{30} + 9S^{20} = (-S^{20} - 3S^{10})^2$$

(g) We have three terms in the expression. This gives us a clue that we are dealing with the square of a sum.

Thus, it must be that:

$$s^{1.25} + s^{2.5} + 0.25 = a^2 + 2ab + b^2 = (a+b)^2$$

$$(s^{5/4})^2 + 2 \cdot s^{5/4} \cdot 0.5 + (0.5)^2 = a^2 + 2ab + b^2$$

This shows that a possible choice is $a = s^{5/4}$ and $b = 0.5$. Finally:

$$s^{1.25} + s^{2.5} + 0.25 = \boxed{\left(\sqrt[4]{s^5} + 0.5\right)^2}$$

Another possible choice is $a = -s^{5/2}$ and $b = -0.5$. That is

$$s^{1.25} + s^{2.5} + 0.25 = \left(-\sqrt[4]{s^5} - 0.5\right)^2$$

(h)

We have four terms in the expression. This gives us a clue that we are dealing with the cube of a sum.

Thus, it must be that:

$$-24\phi^{2/3} + 8\phi + 24\phi^{1/3} - 8 = a^3 + 3a^2b + 3ab^2 + b^3 = (a+b)^3$$

$$(2\phi^{1/3})^3 + 3 \cdot (2\phi^{1/3})^2 \cdot (-2) + 3 \cdot 2\phi^{1/3} \cdot (-2)^2 + (-2)^3 = a^3 + 3a^2b + 3ab^2 + b^3$$

This shows that $a = 2\phi^{1/3}$ and $b = -2$. Finally:

$$-24\phi^{2/3} + 8\phi + 24\phi^{1/3} - 8 = \boxed{\left(2\sqrt[3]{\phi} - 2\right)^3}$$

(i)

We have four terms in the expression. This gives us a clue that we are dealing with the cube of a sum.

Thus, it must be that:

$$150\sqrt[3]{u^5} + 60\sqrt[3]{u^4} + 125u^2 + 8u = a^3 + 3a^2b + 3ab^2 + b^3 = (a+b)^3$$

$$(5\sqrt[3]{u^2})^3 + 3 \cdot (5\sqrt[3]{u^2})^2 \cdot 2\sqrt[3]{u} + 3 \cdot 5\sqrt[3]{u^2} \cdot (2\sqrt[3]{u})^2 + (2\sqrt[3]{u})^3 = a^3 + 3a^2b + 3ab^2 + b^3$$

This shows that $a = 5\sqrt[3]{u^2}$ and $b = 2\sqrt[3]{u}$. Finally:

$$150\sqrt[3]{u^5} + 60\sqrt[3]{u^4} + 125u^2 + 8u = \boxed{\left(5\sqrt[3]{u^2} + 2\sqrt[3]{u}\right)^3}$$

(j) We have three terms in the expression. This gives us a clue that we are dealing with the square of a sum.

Thus, it must be that:

$$16r^5 + 24r^3 + 9r = a^2 + 2ab + b^2 = (a+b)^2$$

$$(4r^{5/2})^2 + 2 \cdot 4r^{5/2} \cdot 3r^{1/2} + (3r^{1/2})^2 = a^2 + 2ab + b^2$$

This shows that a possible choice is $a = 4r^{5/2}$ and $b = 3r^{1/2}$. Finally:

$$16r^5 + 24r^3 + 9r = \boxed{\left(4\sqrt{r^5} + 3\sqrt{r}\right)^2}$$

Another possible choice is $a = -4r^{5/2}$ and $b = -3r^{1/2}$. That is

$$16r^5 + 24r^3 + 9r = \left(-4\sqrt{r^5} - 3\sqrt{r}\right)^2$$

(k)
$$6Q + 16Q^{0.75} + 10Q + 4\sqrt{Q} = 16Q + 16Q^{0.75} + 4\sqrt{Q}$$

We have three terms in the expression. This gives us a clue that we are dealing with the square of a sum.

Thus it must be that:

$$16Q + 16Q^{3/4} + 4\sqrt{Q} = a^2 + 2ab + b^2 = (a+b)^2$$

$$(4\sqrt{Q})^2 + 2 \cdot 4\sqrt{Q} \cdot 2\sqrt[4]{Q} + (2\sqrt[4]{Q})^2 = a^2 + 2ab + b^2$$

This shows that a possible choice is $a = 4\sqrt{Q}$ and $b = 2\sqrt[4]{Q}$. Finally:

$$16Q + 16Q^{0.75} + 4\sqrt{Q} = \boxed{\left(4\sqrt{Q} + 2\sqrt[4]{Q}\right)^2}$$

Another possible choice is $a = -4r^{5/2}$ and $b = -3r^{1/2}$. That is

$$16Q + 16Q^{0.75} + 4\sqrt{Q} = \left(-4\sqrt{Q} - 2\sqrt[4]{Q}\right)^2$$

(l)
$$48f^{23/3} - 12f^{22/3} - 55f^8 + f^7 - (3f^4)^2 = 48f^{23/3} - 12f^{22/3} + f^7 - 64f^8$$

We have four terms in the expression. This gives us a clue that we are dealing with the cube of a sum.

Thus it must be that:

$$48f^{23/3} - 12f^{22/3} + f^7 - 64f^8 = a^3 + 3a^2b + 3ab^2 + b^3 = (a+b)^3$$

$$(f^{7/3})^3 + 3 \cdot (f^{7/3})^2 \cdot (-4f^{8/3}) + 3 \cdot f^{7/3} \cdot (-4f^{8/3})^2 + (-4f^{8/3})^3 = a^3 + 3a^2b + 3ab^2 + b^3$$

This shows that $a = f^{7/3}$ and $b = -4f^{8/3}$. Finally:

$$48f^{23/3} - 12f^{22/3} + f^7 - 64f^8 = \boxed{\left(\sqrt[3]{f^7} - 4\sqrt[3]{f^8}\right)^3}$$

(m)

$$(2 - \sqrt{x})(\sqrt{x} + 2) + 4(x + \sqrt{x}) - 2x = x + 4\sqrt{x} + 4 = (\sqrt{x} + 2)^2$$

$$x + 4\sqrt{x} + 4 = \boxed{(\sqrt{x} + 2)^2}$$

(n)

$$(j^2 + k^2 + jk)(j^6 - 2j^3k^3 + k^6)(j - k)$$

$$= (j - k)(j^2 + +jk + k^2)(j^6 - 2j^3k^3 + k^6) = (j^3 - k^3)(j^3 - k^3)^2 = \boxed{(j^3 - k^3)^3}$$

Exercise 4.7

(a)

$$\frac{1 - s^2}{1 - s} = \frac{(1 - s)(1 + s)}{1 - s} = \boxed{1 + s}$$

Since the original expression contains the division by $1 - s$, we determine that $1 - s \neq 0$ or $s \neq 1$.

(b)

$$(\sqrt{p} + \sqrt{q})^2 - 2\sqrt{p}\sqrt{q} = p + 2\sqrt{pq} + q - 2\sqrt{pq} = \boxed{p + q}$$

Since the original expression contains $\sqrt{p}$ and $\sqrt{q}$, we determine that $p \geq 0$ and $q \geq 0$.

(c)

$$\frac{((W - 2)^2 - (W + 1)^2 + 7W)^2}{W + 3} = \frac{(W^2 - 4W + 4 - (W^2 + 2W + 1) + 7W)^2}{W + 3}$$

$$= \frac{(W + 3)^2}{W + 3} = \boxed{W + 3}$$

Since the original expression contains the division by $W + 3$, we determine that $W + 3 \neq 0$ or $W \neq -3$.

(d)

$$\frac{(\sqrt{\gamma+1}-1)^2 - \gamma - 2}{\sqrt{\gamma+1}} = \frac{\gamma+1-2\sqrt{\gamma+1}+1-\gamma-2}{\sqrt{\gamma+1}} = -\frac{2\sqrt{\gamma+1}}{\sqrt{\gamma+1}} = \boxed{-2}$$

Since the original expression contains the division by $\sqrt{\gamma+1}$, we determine that $\gamma+1 > 0$, or $\gamma > -1$.

(e)

$$\frac{h^3 - g^3}{h - g} + gh = \frac{(h-g)(g^2 + gh + h^2)}{h - g} + gh = g^2 + gh + h^2 + gh = \boxed{(g+h)^2}$$

Since the original expression contains the division by $h - g$ we determine that $h - g \neq 0$ or $h \neq g$.

(f)

$$\frac{K^3 + 12K^2 + 48K + 64}{(K+4)^2} = \frac{(K+4)^3}{(K+4)^2} = \boxed{K+4}$$

Since the original expression contains the division by $(K+4)^2$, we determine that $(K+4)^2 \neq 0$ or $K \neq -4$.

(g)

$$\frac{(\sqrt{z}+\sqrt{z+1})^2 - 2z - 1}{\sqrt{z(z+1)}} = \frac{2z + 2\sqrt{z+1}\sqrt{z}+1-2z-1}{\sqrt{z+1}\sqrt{z}} = \boxed{2}$$

Since the original expression contains the division by $\sqrt{z+1}\sqrt{z}$, we determine that $z+1 > 0$ and $z > 0$. That is $z > -1$ and $z > 0$. Because these two conditions must hold simultaneously, it must be that $z > 0$.

(h)

$$\frac{2t - (\sqrt{t-1}-\sqrt{t+1})^2}{2\sqrt{t-1}\sqrt{t+1}} = \frac{2t - (2t - 2\sqrt{t-1}\sqrt{t+1})}{2\sqrt{t-1}\sqrt{t+1}} = \boxed{1}$$

Since the original expression contains the division by $\sqrt{t-1}\sqrt{t+1}$, we determine that $t-1 > 0$ and $t+1 > 0$. That is $t > 1$ and $t > -1$. Because these two conditions must hold simultaneously, it must be that $t > 1$.

(i)

$$\frac{x^{\frac{9}{4}} + 3x^{\frac{7}{4}} + 3x^{\frac{5}{4}} + x^{\frac{3}{4}}}{x^{\frac{3}{2}} + 2x + x^{\frac{1}{2}}} = \frac{\left(x^{\frac{3}{4}} + x^{\frac{1}{4}}\right)^3}{\left(x^{\frac{3}{4}} + x^{\frac{1}{4}}\right)^2} = \boxed{\sqrt[4]{x^3} + \sqrt[4]{x}}$$

Note that $x^{3/2} + 2x + x^{1/2}$ is equal to $x^{1/2}\left(x^{1/2} + 1\right)^2$ for any $x \geq 0$. But because there is division by $x^{1/2}\left(x^{1/2} + 1\right)^2$, it must be that $x > 0$ and $x^{1/2} \neq -1$. That is, simply $x > 0$.

(j)

$$\left(-\frac{(u+3)(u-3)}{u^2-9}\right)^p = \left(-\frac{u^2-9}{u^2-9}\right)^p = (-1)^p = \boxed{1 \ or \ -1}$$

Since the original expression contains the division by $u^2 - 9$, we determine that $u^2 - 9 \neq 0$ or $u \neq 3$ and $u \neq -3$. Also, p cannot be an irreducible fraction with an even denominator, because this would cause an even root from a negative number.

Exercise 4.8

(a)

$$\frac{(3-n)^2}{(n-6)n+9} = \frac{n^2-6n+9}{n^2-6n+9} = \boxed{1}$$

(b)

$$\left(\frac{\alpha(\alpha+1)+\alpha+4}{\alpha^3-8}\right)^{-1} = \frac{\alpha^3-8}{\alpha(\alpha+1)+\alpha+4} = \frac{(\alpha-2)(\alpha^2+2\alpha+4)}{\alpha^2+2\alpha+4} = \boxed{\alpha-2}$$

(c)

$$\frac{(3-Q^3)(3+Q^3)+2(Q^3-1.5)^2-4.5}{(Q^3-3)^2}$$

$$= \frac{9-Q^6+2(Q^6-3Q^3+2.25)-4.5}{Q^6-6Q^3+9} = \frac{Q^6-6Q^3+9}{Q^6-6Q^3+9} = \boxed{1}$$

(d)

$$\frac{(\sqrt{x}+\sqrt{y})^3}{x+y+2\sqrt{xy}}(\sqrt{x}-\sqrt{y}) = \frac{(\sqrt{x}+\sqrt{y})^2(\sqrt{x}+\sqrt{y})}{x+y+2\sqrt{xy}}(\sqrt{x}-\sqrt{y})$$

$$= \frac{(x+2\sqrt{xy}+y)(\sqrt{x}+\sqrt{y})}{x+2\sqrt{xy}+y}(\sqrt{x}-\sqrt{y}) = (\sqrt{x}+\sqrt{y})(\sqrt{x}-\sqrt{y}) = \boxed{x-y}$$

(e)

$$\frac{k(k-2f)-2f(\sqrt{2f}+\sqrt{k})(\sqrt{k}-\sqrt{2f})}{(k-2f)^2} = \frac{k(k-2f)-2f(k-2f)}{(k-2f)^2}$$

$$= \frac{(k-2f)(k-2f)}{(k-2f)^2} = \frac{(k-2f)^2}{(k-2f)^2} = \boxed{1}$$

(f)

$$\frac{(g-h^4)^3(g+h^4)^2}{h^{16}+g^4-2g^2h^8} = \frac{(g-h^4)^2(g+h^4)^2(g-h^4)}{h^{16}+g^4-2g^2h^8} = \frac{(g^2-h^8)^2(g-h^4)}{(g^2-h^8)^2} = \boxed{g-h^4}$$

(g)

$$\frac{1}{a-b} + \frac{1}{a^2+ab+b^2} - \frac{a+a^2+b^2-b+ab}{a^3-b^3}$$

$$= \frac{a^2+ab+b^2+a-b}{(a-b)(a^2+ab+b^2)} - \frac{a+a^2+b^2-b+ab}{a^3-b^3}$$

$$= \frac{a^2+ab+b^2+a-b}{a^3-b^3} - \frac{a+a^2+b^2-b+ab}{a^3-b^3}$$

$$= \frac{a^2+ab+b^2+a-b-(a^2+ab+b^2+a-b)}{a^3-b^3} = \boxed{0}$$

(h)

$$\frac{(\Delta-2)^4}{\Delta(\Delta(\Delta-6)+12)-8} = \frac{(\Delta-2)^3(\Delta-2)}{\Delta^3-6\Delta^2+12\Delta-88}$$

$$= \frac{(\Delta^3-6\Delta^2+12\Delta-88)(\Delta-2)}{\Delta^3-6\Delta^2+12\Delta-88} = \boxed{\Delta-2}$$

(i)

$$\frac{(p+2)^{-1}}{(\sqrt{p}-\sqrt{2})(\sqrt{p}+\sqrt{2})}\left(\frac{(p-2)^2(p+2)^2}{p^2(48+(p^2-12)p^2)-64}\right)^{-1}$$

$$= \frac{1}{p^2-4} \cdot \frac{p^6-12p^4+48p^2-64}{(p^2-4)^2} = \frac{p^6-12p^4+48p^2-64}{(p^2-4)^3}$$

$$= \frac{p^6-12p^4+48p^2-64}{p^6-12p^4+48p^2-64} = \boxed{1}$$

(j)

$$\frac{(\sqrt{u+v}-\sqrt{u-v})^2+2\sqrt{u^2-v^2}}{(0.25u+1)^2-(0.25u-1)^2}$$

$$= \frac{2u}{(1/16)u^2+(1/2)u+1-((1/16)u^2-(1/2)u+1)} = \frac{2u}{u} = \boxed{2}$$

(k)

$$\frac{s^2(s(s+9)+27)+27s}{2s^2+12s+18} - \frac{s^2((9-s)s-27)+27s}{2s^2-12s+18}$$

$$= \frac{s}{2}\left(\frac{s(s(s+9)+27)+27}{s^2+6s+9} - \frac{s((9-s)s-27)+27}{s^2-6s+9}\right)$$

$$= \frac{s}{2}\left(\frac{s^3+9s^2+27s+27}{s^2+6s+9} - \frac{-s^3+9s^2-27s+27}{s^2-6s+9}\right)$$

$$= \frac{s}{2}\left(\frac{(s+3)^3}{(s+3)^2} + \frac{(s-3)^3}{(s-3)^2}\right) = \frac{s}{2}(s+3+s-3) = \boxed{s^2}$$

(l)

$$\frac{3\sqrt[3]{x^5} - 3\sqrt[3]{x^4} - x^2 + x}{(\sqrt[3]{x} - \sqrt[3]{x^2})(\sqrt[3]{x^4} + \sqrt[3]{x^2} - 2x)} = \frac{(\sqrt[3]{x} - \sqrt[3]{x^2})^3}{(\sqrt[3]{x} - \sqrt[3]{x^2})(\sqrt[3]{x} - \sqrt[3]{x^2})^2} = \frac{(\sqrt[3]{x} - \sqrt[3]{x^2})^3}{(\sqrt[3]{x} - \sqrt[3]{x^2})^3} = \boxed{1}$$

(m)

$$\left(\frac{A^2 B + B^3}{AB} + \frac{A^2 B + B^3}{A^2 + B^2} \right) \frac{A - B}{A^3 - B^3}$$

$$= (A^2 B + B^3) \left(\frac{1}{AB} + \frac{1}{A^2 + B^2} \right) \frac{A - B}{A^3 - B^3}$$

$$= (A^2 B + B^3) \cdot \frac{A^2 + AB + B^2}{AB(A^2 + B^2)} \cdot \frac{A - B}{A^3 - B^3}$$

$$= \frac{(A^2 B + B^3)(A^3 - B^3)}{AB(A^2 + B^2)(A^3 - B^3)} = \frac{B(A^2 + B^2)}{AB(A^2 + B^2)} = \boxed{\dfrac{1}{A}}$$

(n)

$$\frac{c^2 + cd - 2d^2}{c^2 + 4cd + 4d^2} - \frac{3c}{2(c + 2d)} + \frac{1}{2} = \frac{c^2 + cd - 2d^2}{(c + 2d)^2} - \frac{3c}{2(c + 2d)} + \frac{1}{2}$$

$$= \frac{2(c^2 + cd - 2d^2) - 3c(c + 2d)}{2(c + 2d)^2} + \frac{1}{2} = \frac{-c^2 - 4cd - 4d^2}{2(c + 2d)^2} + \frac{1}{2}$$

$$= -\frac{(c + 2d)^2}{2(c + 2d)^2} + \frac{1}{2} = -\frac{1}{2} + \frac{1}{2} = \boxed{0}$$

(o) Denote $A = (x + 4)^5$ for convenience:

$$\frac{A - 5\sqrt{A} + 6}{A - 6\sqrt{A} + 9} - \frac{1}{\sqrt{A} - 3} = \frac{(\sqrt{A} - 3)(\sqrt{A} - 2)}{(\sqrt{A} - 3)^2} - \frac{1}{\sqrt{A} - 3}$$

$$= \frac{\sqrt{A} - 2}{\sqrt{A} - 3} - \frac{1}{\sqrt{A} - 3} = \frac{\sqrt{A} - 3}{\sqrt{A} - 3} = \boxed{1}$$

(p)

$$\frac{(3 + s + t)^2 - (3 - s - t)^2}{s + t} = \frac{(3 + (s + t))^2 - (3 - (s + t))^2}{s + t}$$

$$= \frac{9 + 6(s + t) + (s + t)^2 - (9 - 6(s + t) + (s + t)^2)}{s + t}$$

$$= \frac{9 + 6(s + t) + (s + t)^2 - 9 + 6(s + t) - (s + t)^2}{s + t} = \frac{12(s + t)}{s + t} = \boxed{12}$$

(q)

$$\frac{(2-r-2k)^2 - (2+r-2k)^2}{(1-r-k)^2 - (1+r-k)^2} = \frac{((2-r)-2k)^2 - ((2+r)-2k)^2}{((1-r)-k)^2 - ((1+r)-k)^2}$$

$$= \frac{(2-r)^2 - 4k(2-r) + 4k^2 - ((2+r)^2 - 4k(2+r) + 4k^2)}{(1-r)^2 - 2k(1-r) + k^2 - ((1+r)^2 - 2k(1+r) + k^2)}$$

$$= \frac{8kr - 8r}{4kr - 4r} = \frac{8r(k-1)}{4r(k-1)} = \boxed{2}$$

(r)

$$\frac{(Z+Z^2+1)^3 - (Z-Z^2-1)^3}{(Z^2+1)(Z^4+5Z^2+1)} = \frac{(Z+(Z^2+1))^3 - (Z-(Z^2+1))^3}{(Z^2+1)(Z^4+5Z^2+1)}$$

$$= \frac{2Z^6 + 12Z^4 + 12Z^2 + 2}{Z^6 + 6Z^4 + 6Z^2 + 1} = \frac{2(Z^6 + 6Z^4 + 6Z^2 + 1)}{Z^6 + 6Z^4 + 6Z^2 + 1} = \boxed{2}$$

(s)

$$\frac{(1+f)(1-f)(1+2f) - (1-f)(1-2f)(1+2f)}{f(f(2f-1)-1)} = \frac{(1-f^2)(1+2f) - (1-f)(1-4f^2)}{2f^3 - f^2 - f}$$

$$= \frac{(1-f^2)(1+2f) - (1-f)(1-4f^2)}{2f^3 - f^2 - f} = \frac{-6f^3 + 3f^2 + 3f}{2f^3 - f^2 - f}$$

$$= -\frac{3(2f^3 - f^2 - f)}{2f^3 - f^2 - f} = \boxed{-3}$$

(t)

$$\left(\frac{q^3}{q^3-8} + \frac{8(q-2)^{-2}}{q^2+2q+4}\right) \frac{q((q-2)q^2-8) + 16}{q^3(q-2) + 8}$$

$$= \left(\frac{q^3}{q^3-8} + \frac{8}{(q-2)^2(q^2+2q+4)}\right) \frac{q^4 - 2q^3 - 8q + 16}{q^4 - 2q^3 + 8}$$

$$= \left(\frac{q^3}{q^3-8} + \frac{8}{(q-2)(q^3-8)}\right) \frac{(q-2)(q^3-8)}{q^4 - 2q^3 + 8}$$

$$= \frac{q^3(q-2) + 8}{(q-2)(q^3-8)} \cdot \frac{(q-2)(q^3-8)}{q^4 - 2q^3 + 8} = \frac{q^4 - 2q^3 + 8}{(q-2)(q^3-8)} \cdot \frac{(q-2)(q^3-8)}{q^4 - 2q^3 + 8} = \boxed{1}$$

(u)

$$\frac{(w + w^4 + w^5 + w^8)^3}{w^{11}(w^4 + 3) + w^3(3w^4 + 1)}(w + 1)^{-2}(w^2 - w + 1)^{-3}$$

$$= \frac{(w(1 + w^3) + w^5(1 + w^3))^3}{w^{15} + 3w^{11} + 3w^7 + w^3}(w + 1)^{-2}(w^2 - w + 1)^{-3}$$

$$= \frac{(w + w^5)^3(1 + w^3)^3}{(w^5 + w)^3} \cdot \frac{w + 1}{(w + 1)^3(w^2 - w + 1)^3}$$

$$= \frac{(1 + w^3)^3(w + 1)}{(1 + w^3)^3} = \boxed{w + 1}$$

Exercise 4.9

$$\frac{(x^2 - 2x + 4)(\sqrt{x + 3} + \sqrt{x + 1})^2 - 2(x^3 + 8)}{(x - 1)^2\sqrt{x^2 + 4x + 3} + 3\sqrt{x^2 + 4x + 3}}$$

$$= \frac{(x^2 - 2x + 4)(2x + 2\sqrt{x + 1}\sqrt{x + 3} + 4) - 2(x + 2)(x^2 - 2x + 4)}{((x - 1)^2 + 3)\sqrt{x^2 + 4x + 3}}$$

$$= \frac{(x^2 - 2x + 4)(2x + 2\sqrt{x + 1}\sqrt{x + 3} + 4 - 2x - 4)}{(x^2 - 2x + 4)\sqrt{x^2 + 4x + 3}}$$

$$= \frac{2\sqrt{x + 1}\sqrt{x + 3}}{\sqrt{x^2 + 4x + 3}} = \frac{2\sqrt{x + 1}\sqrt{x + 3}}{\sqrt{(x + 1)(x + 3)}} = \boxed{2}$$

Exercise 4.10

$$\left(\frac{(a^3 - 64)((a + 4)^3 - 12a(a + 4))}{a^3 + 64}\right)^3 \frac{(a^2 - 8\sqrt{a})^{-1}}{a^5 + 8a^3\sqrt{a} - 64a^2 - 512\sqrt{a}}$$

$$= \left(\frac{(a^3 - 64)(a + 4)((a + 4)^2 - 12a)}{a^3 + 64}\right)^3 \frac{(a^2 - 8\sqrt{a})^{-1}}{a^3(a^2 + 8\sqrt{a}) - 64(a^2 + 8\sqrt{a})}$$

$$\left(\frac{(a^3 - 64)(a + 4)(a^2 - 4a + 16)}{a^3 + 64}\right)^3 \frac{(a^2 - 8\sqrt{a})^{-1}}{(a^2 + 8\sqrt{a})(a^3 - 64)}$$

$$= \left(\frac{(a^3 - 64)(a^3 + 64)}{a^3 + 64}\right)^3 \frac{(a^2 - 8\sqrt{a})^{-1}}{(a^2 + 8\sqrt{a})(a^3 - 64)}$$

$$= (a^3 - 64)^3 \frac{1}{(a^2 - 8\sqrt{a})(a^2 + 8\sqrt{a})(a^3 - 64)}$$

$$= \frac{(a^3 - 64)^2}{a^4 - 64a} = \frac{(a^3 - 64)^2}{a(a^3 - 64)} = \boxed{\frac{a^3 - 64}{a}}$$

Evaluating the expression for $a = 2$:

$$\frac{a^3 - 64}{a} = \frac{2^3 - 64}{2} = \boxed{-28}$$

Exercise 4.11

1. First, simplify the numerator:

$$\frac{(1+t+t^2+t^3)^2(t^3+t)}{(t+1)^2} - ((t^2+3)t^2+3)t^3$$

$$= \frac{((1+t)+t^2(t+1))^2 t(t^2+1)}{(t+1)^2} - ((t^2+3)t^2+3)t^3$$

$$= \frac{(1+t)^2(1+t^2)^2 t(t^2+1)}{(t+1)^2} - ((t^2+3)t^2+3)t^3$$

$$= t(t^2+1)^3 - (t^7+3t^5+3t^3) = t$$

2. Next, simplify the denominator:

$$(1-t+t^2-t^3)^3(t-1)^{-2} - (t^3-t^2)(t^2(-t^2-3)-3)$$

$$= \frac{(1-t+t^2-t^3)^3}{(t-1)^2} - (t^3-t^2)(t^2(-t^2-3)-3)$$

$$= \frac{((1-t)+t^2(1-t))^3}{(t-1)^2} - t^2(t-1)(t^2(-t^2-3)-3)$$

$$= \frac{(1-t)^3(1+t^2)^3}{(t-1)^2} - t^2(t-1)(t^2(-t^2-3)-3)$$

$$= (1-t)(1+t^2)^3 - t^2(t-1)(t^2(-t^2-3)-3)$$

$$= (1-t)((1+t^2)^3 + t^2(t^2(-t^2-3)-3))$$

$$= (1-t)((1+t^2)^3 - t^6 - 3t^4 - 3t^2) = 1-t$$

3. Thus, our expression becomes:

$$\boxed{\frac{t}{1-t}}$$

4. Evaluate the expression for $t = 0.5$:

$$\frac{t}{1-t} = \frac{0.5}{1-0.5} = \boxed{1}$$

Exercise 4.12

$$\frac{\left(\sqrt{-x^2 - 2x} + \sqrt{x^2 + 2x + 2} - \sqrt{2}\right)\left(\sqrt{1 - (1+x)^2} + \sqrt{x^2 + 2x + 2} + \sqrt{2}\right)}{\sqrt{-(x+1)^5 + x + 1}}$$

$$= \frac{\left(\sqrt{1 - (1+x)^2} + \sqrt{1 + (1+x)^2} - \sqrt{2}\right)\left(\sqrt{1 - (1+x)^2} + \sqrt{1 + (1+x)^2} + \sqrt{2}\right)}{\sqrt{1 + x}\sqrt{1 - (x+1)^4}}$$

$$= \frac{\left(\sqrt{1 - (1+x)^2} + \sqrt{1 + (1+x)^2}\right)^2 - 2}{\sqrt{1 + x}\sqrt{1 - (x+1)^4}} = \frac{2\sqrt{1 - (x+1)^2}\sqrt{1 + (x+1)^2} + 2 - 2}{\sqrt{1 + x}\sqrt{1 - (x+1)^4}}$$

$$= \frac{2\sqrt{1 - (x+1)^4}}{\sqrt{1 + x}\sqrt{1 - (x+1)^4}} = \boxed{\frac{2}{\sqrt{1 + x}}}$$

Evaluating the expression for $t = -\frac{3}{4}$:

$$\frac{2}{\sqrt{1 + x}} = \frac{2}{\sqrt{1 - 0.75}} = \frac{2}{\sqrt{0.25}} = \boxed{4}$$

Exercise 4.13

$$\frac{(2u - v)^3 - u^2(6u - 13v)}{u^3 - uv^2} - \frac{(u - v)^2}{u^2 + uv} - \frac{(u + v)^2}{u^2 - uv}$$

$$= \frac{(2u - v)^3 - u^2(6u - 13v)}{u(u^2 - v^2)} - \frac{(u - v)^2}{u(u + v)} - \frac{(u + v)^2}{u(u - v)}$$

$$= \frac{(2u - v)^3 - u^2(6u - 13v) - (u - v)^3 - (u + v)^3}{u(u^2 - v^2)}$$

$$= \frac{u^2 v - v^3}{u(u^2 - v^2)} = \frac{v(u^2 - v^2)}{u(u^2 - v^2)} = \boxed{\frac{v}{u}}$$

Evaluating the expression for $u = 4$ and $v = 8$:

$$\frac{v}{u} = \frac{8}{4} = \boxed{2}$$

Exercise 4.14

$$\frac{(z+t)^4 - (z^2+t^2)^2}{(t^3-z^3)((t-z)^2-(t+z)^2)} - \frac{1}{z+t} + \frac{3t+z}{t^2-z^2}$$

$$= \frac{(z+t)^2(z+t)^2 - (z^2+t^2)^2}{(t^3-z^3)(-4tz)} + \frac{z-t+3t+z}{t^2-z^2}$$

$$= \frac{t^4+4t^3z+6t^2z^2+4tz^3+z^4 - (t^4+2t^2z^2+z^4)}{(t^3-z^3)(-4tz)} + \frac{2z+2t}{t^2-z^2}$$

$$= \frac{4tz^3+4t^2z^2+4t^3z}{(t^3-z^3)(-4tz)} + \frac{2}{t-z} = \frac{4tz(t^2+tz+z^2)}{(t-z)(t^2+tz+z^2)(-4tz)} + \frac{2}{t-z}$$

$$= -\frac{1}{t-z} + \frac{2}{t-z} = \boxed{\frac{1}{t-z}}$$

Evaluating the expression for $t = 0.5$ and $z = 0.25$:

$$\frac{1}{t-z} = \frac{1}{0.5-0.25} = \frac{1}{0.25} = \boxed{4}$$

Exercise 4.15

$$\frac{0.25}{x^2+x+1}\left(\sqrt{-\frac{x^3-1}{1-x}} + \sqrt{-\frac{x^4+x^3-x-1}{1-x^2}}\right)^2$$

$$= \frac{0.25}{x^2+x+1}\left(\sqrt{-\frac{x^3-1}{1-x}} + \sqrt{-\frac{(1+x)(x^3-1)}{1-x^2}}\right)^2$$

$$= \frac{0.25(x^3-1)}{x^2+x+1}\left(\sqrt{-\frac{1}{1-x}} + \sqrt{-\frac{1+x}{1-x^2}}\right)^2 = \frac{0.25(x^3-1)}{x^2+x+1} \cdot \frac{4}{x-1}$$

$$= \frac{0.25 \cdot 4(x^3-1)}{x^3-1} = \boxed{1}$$

Exercise 4.16

$$\left(\frac{1}{t-1} + \frac{t+1}{(t-1)^2} + \frac{t^2+2t+1}{(t-1)^3} + \frac{t^3+3t^2+3t+1}{(t-1)^4}\right) \cdot \frac{(t-1)^4}{t^2+1}$$

$$= \left(\frac{1}{t-1} + \frac{t+1}{(t-1)^2} + \frac{(t+1)^2}{(t-1)^3} + \frac{(t+1)^3}{(t-1)^4}\right) \cdot \frac{(t-1)^4}{t^2+1}$$

Let's denote $A = t+1$ and $B = t-1$ to make the expression more transparent:

$$\left(\frac{1}{B} + \frac{A}{B^2} + \frac{A^2}{B^3} + \frac{A^3}{B^4}\right) \cdot \frac{B^4}{t^2+1} = \frac{B^3 + AB^2 + A^2B + A^3}{B^4} \cdot \frac{B^4}{t^2+1}$$

$$= \frac{B^3 + AB^2 + A^2B + A^3}{t^2+1} = \frac{B^3 + 3AB^2 + 3A^2B + A^3 - 2AB^2 - 2A^2B}{t^2+1}$$

$$= \frac{(A+B)^3 - 2AB(A+B)}{t^2+1}$$

We can return to our variable t:

$$\frac{(2t)^3 - 2(t^2-1)(2t)}{t^2+1} = \frac{8t^3 - 4t^3 + 4t}{t^2+1} = \frac{4t^3 + 4t}{t^2+1} = \frac{4t(t^2+1)}{t^2+1} = \boxed{4t}$$

Exercise 4.17

$$\left(A^B A^C\right)^D = A^{(B+C)D}$$

$$A = 1 - \left(1 + \sqrt{2s}\right)\left(1 - \sqrt{2s}\right)(1+2s) = 1 - (1-2s)(1+2s) = 4s^2$$

$$B + C = \frac{\sqrt{s+\sqrt{s}}}{\sqrt{s-\sqrt{s}}} + \frac{\sqrt{s-\sqrt{s}}}{\sqrt{s+\sqrt{s}}} = \frac{s+\sqrt{s}+s-\sqrt{s}}{\sqrt{s^2-s}} = \frac{2s}{\sqrt{s}\sqrt{s-1}}$$

$$(B+C)D = \frac{2s}{\sqrt{s}\sqrt{s-1}} \cdot \frac{\sqrt{s-1}}{4\sqrt{s}} = \frac{1}{2}$$

$$A^{(B+C)D} = \sqrt{4s^2} = \boxed{2s}$$

Note that the answer is not $2|s|$, because it must be that $s > 1$.

Exercise 4.18

$$((\alpha^x \beta^x)^y)^{-2z} = (\alpha\beta)^{-2xyz}$$

$$\alpha\beta = \sqrt[4]{\frac{F+D}{F^2-D^2}} \cdot \sqrt[4]{\frac{1}{F-D}} = \sqrt{\frac{1}{F-D}}$$

$$-2xyz = -2 \cdot \frac{F^3-D^3}{(F^{1.5}-D^{1.5})(\sqrt{FD}+D)} \cdot \frac{D^{2.5}-DF^{1.5}}{F^2+DF+D^2} \cdot \frac{F+\sqrt{DF}+D}{D^2-F\sqrt{DF}}$$

$$= -2 \cdot \frac{(F-D)(F^2+DF+D^2)}{(F^{1.5}-D^{1.5})(\sqrt{FD}+D)} \cdot \frac{D(D^{1.5}-F^{1.5})}{F^2+DF+D^2} \cdot \frac{F+\sqrt{DF}+D}{D^2-F\sqrt{DF}}$$

$$= 2 \cdot \frac{F-D}{\sqrt{FD}+D} \cdot \frac{D}{1} \cdot \frac{F+\sqrt{DF}+D}{D^2-F\sqrt{DF}}$$

$$= 2 \cdot \frac{F-D}{\sqrt{D}\left(\sqrt{F}+\sqrt{D}\right)} \cdot \frac{D}{1} \cdot \frac{F+\sqrt{DF}+D}{D^2-F\sqrt{DF}}$$

$$= \frac{2\left(\sqrt{F}-\sqrt{D}\right)\left(F+\sqrt{DF}+D\right)\sqrt{D}}{D^2-F\sqrt{DF}} = \frac{2\left(\sqrt{F^3}-\sqrt{D^3}\right)\sqrt{D}}{D^2-F\sqrt{DF}}$$

$$= \frac{2\sqrt{D}\left(\sqrt{F^3}-\sqrt{D^3}\right)}{\sqrt{D}\left(\sqrt{D^3}-\sqrt{F^3}\right)} = -2$$

$$(\alpha\beta)^{-2xyz} = \left(\sqrt{\frac{1}{F-D}}\right)^{-2} = \boxed{F-D}$$

Evaluating the expression for $D = 3$ and $F = 4$:

$$F - D = 4 - 3 = \boxed{1}$$

Exercise 4.19

$$\frac{s}{w} = \frac{4ab(3a^2-b^2)}{(\sqrt{12}a+2b)(\sqrt{3}a-b)(3a^2+b^2)} \cdot \frac{(a+b)^3-(a-b)^3}{(a+b)^2-(a-b)^2}$$

$$= \frac{4ab(3a^2-b^2)}{2(\sqrt{3}a+b)(\sqrt{3}a-b)(3a^2+b^2)} \cdot \frac{6a^2b+2b^3}{4ab}$$

$$= \frac{3a^2-b^2}{2(3a^2-b^2)(3a^2+b^2)} \cdot \frac{6a^2b+2b^3}{1} = b$$

$$\frac{q}{p} = \left(\frac{1}{a} - \frac{1 - a^3 b^3}{a(1 + ab)^2 - a^2 b}\right) \div \left(\left(\frac{1}{(a + 1)^3} - \frac{1}{(a^2 - 1)^3}\right) \cdot \frac{a^6 - 3a^4 + 3a^2 - 1}{a^3 - 3a^2 + 3a - 2}\right)$$

$$= \left(\frac{1}{a} - \frac{1 - a^3 b^3}{a(a^2 b^2 + ab + 1)}\right) \div \left(\left(\frac{1}{(a + 1)^3} - \frac{1}{(a - 1)^3 (a + 1)^3}\right) \cdot \frac{(a^2 - 1)^3}{(a - 1)^3 - 1}\right)$$

$$= \left(\frac{a^2 b^2 + ab + 1 - 1 + a^3 b^3}{a^3 b^2 + a^2 b + a}\right) \div \left(\frac{(a - 1)^3 - 1}{(a - 1)^3 (a + 1)^3} \cdot \frac{(a^2 - 1)^3}{(a - 1)^3 - 1}\right)$$

$$= \frac{b(a^3 b^2 + a^2 b + a)}{a^3 b^2 + a^2 b + a} \div 1 = b$$

Thus:

$$\sqrt[w]{\xi^s} \sqrt[p]{\phi^q} = \xi^b \phi^b = (\xi \phi)^b$$

Let's now find $\xi \phi$:

$$\xi = \left(\frac{(a + a^2 + 3)^2}{a(1 + a)} - \frac{a^4 + 2a^3 + a^2 + 27(a + a^2)^{-1}}{(a + 0.5)^2 + 2.75}\right)$$

$$= \left(\frac{(a + a^2 + 3)^2}{a + a^2} - \frac{(a + a^2)^2 + 27(a + a^2)^{-1}}{a + a^2 + 3}\right)$$

$$= \{denote\ A = a + a^2\} = \left(\frac{(A + 3)^2}{A} - \frac{A^2 + 27A^{-1}}{A + 3}\right)$$

$$= \frac{(A + 3)^3 - A^3 - 27}{A(A + 3)} = \frac{9A^2 + 27A}{A(A + 3)} = \frac{9A(A + 3)}{A(A + 3)} = 9$$

$$\phi = \frac{(b + 1)^4 - b(b - 2)^3 - (b - 2)^3}{b^3 + 1} = \frac{(b + 1)^3 (b + 1) - (b - 2)^3 (b + 1)}{b^3 + 1}$$

$$= \frac{(b + 1)((b + 1)^3 - (b - 2)^3)}{b^3 + 1} = \frac{(b + 1)(9b^2 - 9b + 9)}{b^3 + 1} = \frac{9(b + 1)(b^2 - b + 1)}{(b + 1)(b^2 - b + 1)} = 9$$

We have: $\xi \phi = 81$. The last task is to simplify v:

$$v = \left(3 + 2\sqrt{b}\right)\left(a + 2\sqrt{b}\right) - 2(a + 3)\sqrt{b} - 3a$$

$$= 2a\sqrt{b} + 3a + 4b + 6\sqrt{b} - 2a\sqrt{b} - 3a - 6\sqrt{b} = 4b$$

Finally:

$$\sqrt[v]{\sqrt[w]{\xi^s} \sqrt[p]{\phi^q}} = \sqrt[4b]{81^b} = 81^{\frac{b}{4b}} = 81^{\frac{1}{4}} = \boxed{3}$$

Exercise 4.20

$$(x-y)^{\frac{1}{256}} \; {}^{256}\!\!\sqrt{\frac{(x+y)(x^2+y^2)(x^4+y^4)(x^8+y^8)\cdots(x^{128}+y^{128})}{y^{256}} + \frac{1}{x-y}}$$

$$= (x-y)^{\frac{1}{256}} \; {}^{256}\!\!\sqrt{\frac{(x-y)(x+y)(x^2+y^2)(x^4+y^4)(x^8+y^8)\cdots(x^{128}+y^{128}) + y^{256}}{y^{256}(x-y)}}$$

$$= {}^{256}\!\!\sqrt{x-y} \; {}^{256}\!\!\sqrt{\frac{x^{256}-y^{256}+y^{256}}{y^{256}(x-y)}} = {}^{256}\!\!\sqrt{x-y} \; {}^{256}\!\!\sqrt{\frac{1}{x-y}} \; {}^{256}\!\!\sqrt{\frac{x^{256}}{y^{256}}}$$

$$= {}^{256}\!\!\sqrt{\left(\frac{x}{y}\right)^{265}} = \boxed{\left|\frac{x}{y}\right|}$$

We have ${}^{256}\!\!\sqrt{x-y}$, which means that $x - y \geq 0$. But we also have the division by $x - y$, which means that $x - y \neq 0$. Thus $x - y > 0$, or $x > y$. We also have the division by y^{256}, which gives $y \neq 0$.

Exercise 4.21

(a) Let's write $A - B$ in the following form:

$$A - B = (1+t)^2 - (1-t)^2 + (2+t)^2 - (2-t)^2 + \cdots + (n+t)^2 - (n-t)^2$$

Note that for any n:
$$(n+t)^2 - (n-t)^2 = 4nt$$

Thus:

$$A - B = 4t + 8t + 12t + \cdots 4nt = 4t(1 + 2 + 3 + \cdots + n)$$

$$= 4t\frac{n(n+1)}{2} = \boxed{2n(n+1)t}$$

(b)

$$2n(n+1)t = 2 \cdot 10(10+1)2 = \boxed{440}$$

Exercise 4.22

(a)

$$\frac{n^3}{3} - \frac{(n-1)^3}{3} + n - \frac{1}{3} = \frac{n^3 - (n-1)^3 + 3n - 1}{3}$$

$$= \frac{n^3 - (n^3 - 3n^2 + 3n - 1) + 3n - 1}{3} = \frac{3n^2}{3} = \boxed{n^2}$$

(b)

$$1^2 + 2^2 + 3^2 + \cdots + n^2$$

$$= \left(\frac{1^3}{3} - \frac{0^3}{3} + 1 - \frac{1}{3}\right) + \left(\frac{2^3}{3} - \frac{1^3}{3} + 2 - \frac{1}{3}\right) + \left(\frac{3^3}{3} - \frac{2^3}{3} + 3 - \frac{1}{3}\right) +$$

$$+ \cdots + \left(\frac{n^3}{3} - \frac{(n-1)^3}{3} + n - \frac{1}{3}\right)$$

$$= \left(\frac{1^3}{3} - \frac{1^3}{3} + \frac{2^3}{3} - \frac{2^3}{3} + \frac{3^3}{3} - \frac{3^3}{3} + \cdots + \frac{(n-1)^3}{3} - \frac{(n-1)^3}{3}\right) +$$

$$+ \frac{n^3}{3} + (1 + 2 + 3 + \cdots + n) - \frac{n}{3}$$

That is, using Exercise 3.20(b):

$$1^2 + 2^2 + 3^2 + \cdots + n^2 = \frac{n^3}{3} + (1 + 2 + 3 + \cdots + n) - \frac{n}{3} = \frac{n^3}{3} + \frac{n(n+1)}{2} - \frac{n}{3}$$

$$= \frac{2n^3 + 3n^2 + n}{6} = \frac{n(n+1)(2n+1)}{6}$$

(c)

$$1^2 + 2^2 + 3^2 + \cdots + 12^2 = \frac{12(12+1)(2 \cdot 12 + 1)}{6} = 2 \cdot 25 \cdot 13 = \boxed{650}$$

Exercise 4.23

(a)

$$\frac{n^2(n+1)^2}{4} - \frac{(n-1)^2 n^2}{4} = \frac{n^2(n^2 + 2n + 1) - (n^2 - 2n + 1)n^2}{4} = \frac{4n^3}{4} = \boxed{n^3}$$

(b)

$$1^3 + 2^3 + 3^3 + \cdots + n^3$$

$$= \left(\frac{1^2 \cdot 2^2}{4} - \frac{0^2 \cdot 1^2}{4} \right) + \left(\frac{2^2 \cdot 3^2}{4} - \frac{1^2 \cdot 2^2}{4} \right) + \left(\frac{3^2 \cdot 4^2}{4} - \frac{2^2 \cdot 3^2}{4} \right) +$$

$$+ \cdots + \left(\frac{n^2(n+1)^2}{4} - \frac{(n-1)^2 n^2}{4} \right)$$

$$= \left(\frac{1^2 \cdot 2^2}{4} - \frac{1^2 \cdot 2^2}{4} + \frac{2^2 \cdot 3^2}{4} - \frac{2^2 \cdot 3^2}{4} + \cdots + \frac{(n-1)^2 \cdot n^2}{4} - \frac{(n-1)^2 \cdot n^2}{4} \right) +$$

$$+ \frac{n^2(n+1)^2}{4} = \frac{n^2(n+1)^2}{4} = (1 + 2 + 3 + \cdots + n)^2$$

(c)

$$1^3 + 2^3 + 3^3 + \cdots + 10^3 = \left(\frac{n(n+1)}{2} \right)^2 = \left(\frac{10 \cdot 11}{2} \right)^2 = 55^2 = \boxed{3025}$$

Exercise 4.24

(a) Similarly to the solution of Problem 4.21 and using Problem 4.22(b):

$$(n+d)^3 - (n-d)^3 = 2d^3 + 6dn^2$$

$$A - B = 2d^3 + 6d \cdot 1^2 + 2d^3 + 6d \cdot 2^2 + 2d^3 + 6d \cdot 3^2 + \cdots + 2d^3 + 6d \cdot n^2$$

$$= 2d^3 n + 6d(1^2 + 2^2 + 3^2 + \cdots + n^2)$$

$$= 2d^3 n + 6d \frac{n(2n+1)(n+1)}{6} = \boxed{2d^3 n + dn(n+1)(2n+1)}$$

(b)

$$2d^3 n + dn(n+1)(2n+1) = 2 \cdot 2^3 \cdot 10 + 2 \cdot 10(10+1)(2 \cdot 10+1) = \boxed{4780}$$

Exercise 4.25

$$(1 + nr)^3 - (1 - nr)^3 = 2n^3 r^3 + 6nr$$

$$A - B = 2 \cdot 1^3 r^3 + 6 \cdot 1r + 2 \cdot 2^3 r^3 + 6 \cdot 2r + 2 \cdot 3^3 r^3 + 6 \cdot 3r + \cdots + 2 \cdot n^3 r^3 + 6 \cdot nr$$

$$= 2r^3(1^3 + 2^3 + 3^3 + \cdots + n^3) + 6r(1 + 2 + 3 + \cdots + n)$$

$$= 2r^3 \left(\frac{n(n+1)}{2} \right)^2 + 6r \frac{n(n+1)}{2} = \frac{1}{2} n^2(n+1)^2 r^3 + 3n(n+1)r$$

$$\frac{A - B}{C} = \frac{1}{2} \cdot \frac{n^2(n+1)^2 r^3 + 6n(n+1)r}{n^4 r^3 + 2n^3 r^3 + n^2 r^3 + 6n^2 r + 6nr}$$

$$= \frac{1}{2} \cdot \frac{n^4 r^3 + 2n^3 r^3 + n^2 r^3 + 6n^2 r + 6nr}{n^4 r^3 + 2n^3 r^3 + n^2 r^3 + 6n^2 r + 6nr} = \boxed{0.5}$$

Solutions for Chapter 5 Exercises

Exercise 5.1

(a)
$$x^2 - 3x + 2 = x^2 - 2x - x + 2 = x(x-1) - 2(x-1) = \boxed{(x-1)(x-2)}$$

(b)
$$x^2 - x - 12 = x^2 - 4x + 3x - 12 = x(x-4) + 3(x-4) = \boxed{(x+3)(x-4)}$$

(c)
$$6x^2 - 7x - 5 = 6x^2 - 10x + 3x - 5 = 2x(3x-5) + (3x-5) = \boxed{(2x+1)(3x-5)}$$

(d)
$$x^4 - 3x^2 + 2 = x^4 - x^2 - 2x^2 + 2 = x^2(x^2-1) - 2(x^2-1) = \boxed{(x^2-2)(x^2-1)}$$

(e)
$$2x^3 - 6x^2 - x + 3 = 2x^2(x-3) - (x-3) = \boxed{(2x^2-1)(x-3)}$$

(f)
$$x^6 + 3x^4 + x^2 + 3 = x^2(x^4+1) + 3(x^4+1) = \boxed{(x^2+3)(x^4+1)}$$

(g)
$$x^4 - x^3 - 2x^2 = x^4 - 2x^3 + x^3 - 2x^2 = x^2(x^2-2x) + x(x^2-2x) = \boxed{(x^2+x)(x^2-2x)}$$

(h)
$$x^4 - 2x^3 + 5x^2 - 10x = x^2(x^2+5) - 2x(x^2+5) = \boxed{(x^2-2x)(x^2+5)}$$

(i)
$$x^2 + xy - 2y^2 = x^2 + 2xy - xy - 2y^2 = x(x-y) + 2y(x-y) = \boxed{(x+2y)(x-y)}$$

(j)
$$x^5 + 2x^3y + 2x^2 + 4y = x^2(x^3+2) + 2y(x^3+2) = \boxed{(x^2+2y)(x^3+2)}$$

Exercise 5.2

(a)

$$\frac{x^4 - 16}{x + 2} = \frac{(x^2 - 4)(x^2 + 4)}{x + 2} = \frac{(x - 2)(x + 2)(x^2 + 4)}{x + 2}$$

$$= (x - 2)(x^2 + 4) = \boxed{x^3 - 2x^2 + 4x - 8}$$

$$
\begin{array}{r}
x^3 - 2x^2 + 4x\ - 8 \\
x + 2 \overline{)\ x^4 \qquad\qquad\qquad - 16} \\
\underline{-x^4 - 2x^3} \\
-2x^3 \\
\underline{2x^3 + 4x^2} \\
4x^2 \\
\underline{-4x^2 - 8x} \\
-8x - 16 \\
\underline{8x + 16} \\
0
\end{array}
$$

(b)

$$\frac{9x^2 + 24x + 16}{3x + 4} = \frac{(3x + 4)^2}{3x + 4} = \boxed{3x + 4}$$

$$
\begin{array}{r}
3x\ + 4 \\
3x + 4 \overline{)\ 9x^2 + 24x + 16} \\
\underline{-9x^2 - 12x} \\
12x + 16 \\
\underline{-12x - 16} \\
0
\end{array}
$$

(c)

$$\frac{9x^6 + 18x^4 + 9x^2}{x^3 + x} = \frac{9(x^3 + x)^2}{x^3 + x} = \boxed{9x^3 + 9x}$$

$$
\begin{array}{r}
9x^3 \qquad\quad + 9x \\
x^3 + x \overline{\smash{)}\ 9x^6 + 18x^4 \qquad + 9x^2} \\
\underline{-\,9x^6\ -\ 9x^4} \\
9x^4 \qquad + 9x^2 \\
\underline{-\,9x^4 \qquad -\ 9x^2} \\
0
\end{array}
$$

(d)

$$
\frac{4x^6 - 8x^5 + 4x^4}{x^2 - 2x + 1} = \frac{4x^4(x^2 - 2x + 1)}{x^2 - 2x + 1} = \boxed{4x^4}
$$

$$
\begin{array}{r}
4x^4 \\
x^2 - 2x + 1 \overline{\smash{)}\ 4x^6 - 8x^5 + 4x^4} \\
\underline{-\,4x^6 + 8x^5 - 4x^4} \\
0
\end{array}
$$

(e)

$$
\frac{x^3 - 9x^2 + 27x - 27}{x - 3} = \frac{(x - 3)^3}{x - 3} = (x - 3)^2 = \boxed{x^2 - 6x + 9}
$$

$$
\begin{array}{r}
x^2\ -\ 6x\ +9 \\
x - 3 \overline{\smash{)}\ x^3 - 9x^2 + 27x - 27} \\
\underline{-\,x^3 + 3x^2} \\
-6x^2 + 27x \\
\underline{6x^2 - 18x} \\
9x - 27 \\
\underline{-\,9x + 27} \\
0
\end{array}
$$

(f)

$$
\frac{x^3 - 8}{x - 2} = \frac{(x - 2)(x^2 + 2x + 4)}{x - 2} = \boxed{x^2 + 2x + 4}
$$

$$
\begin{array}{r}
x^2 + 2x + 4 \\
x - 2 \overline{)\ x^3 \qquad\qquad -\ 8} \\
\underline{-\ x^3 + 2x^2} \\
2x^2 \\
\underline{-\ 2x^2 + 4x} \\
4x - 8 \\
\underline{-\ 4x + 8} \\
0
\end{array}
$$

(g)

$$
\frac{3x^2 + 9x - 12}{x + 4} = \frac{3(x - 1)(x + 4)}{x + 4} = \boxed{3x - 3}
$$

$$
\begin{array}{r}
3x \quad -\ 3 \\
x + 4 \overline{)\ 3x^2 \ +9x - 12} \\
\underline{-\ 3x^2 - 12x} \\
-\ 3x - 12 \\
\underline{3x + 12} \\
0
\end{array}
$$

(h)

$$
\frac{6x^2 + 11x + 3}{3x + 1} = \frac{(3x + 1)(2x + 3)}{3x + 1} = \boxed{2x + 3}
$$

$$
\begin{array}{r}
2x + 3 \\
3x + 1 \overline{)\ 6x^2 + 11x + 3} \\
\underline{-\ 6x^2 \ -\ 2x} \\
9x + 3 \\
\underline{-\ 9x - 3} \\
0
\end{array}
$$

(i)

$$
\frac{x^3 y^6 - x^3 y^3}{y^2 + y + 1} = \frac{x^3 y^3 (y^3 - 1)}{y^2 + y + 1} = \frac{x^3 y^3 (y - 1)(y^2 + y + 1)}{y^2 + y + 1} = x^3 y^3 (y - 1) = \boxed{x^3 y^4 - x^3 y^3}
$$

(j)

$$
\frac{x^3 y^3 + 3x^2 y^3 + 3xy^3 + y^3}{xy + y} = \frac{(xy + y)^3}{xy + y} = (xy + y)^2 = \boxed{x^2 y^2 + 2xy^2 + y^2}
$$

(k)

$$\frac{x^2 + 2xy + 6x + y^2 + 6y + 9}{x + y + 3} = \frac{(x+y)^2 + 6(x+y) + 9}{x + y + 3} = \frac{(x+y+3)^2}{x + y + 3} = \boxed{x + y + 3}$$

(l)

$$\frac{x^3 + 3x^2y - x^2 + 3xy^2 - 2xy + y^3 - y^2}{x + y - 1} = \frac{(x+y)^3 - (x+y)^2}{x + y - 1}$$

$$= \frac{(x+y)^2(x+y-1)}{x + y - 1} = \boxed{x^2 + 2xy + y^2}$$

Exercise 5.3

(a)

$$3x^2 + 6x + 6 = 3\left(x^2 + 2x + 2\right) = 3\left(x^2 + 2x + 1 + 1\right) = \boxed{3(x+1)^2 + 3}$$

(b)

$$2x^2 + 8x + 7 = 2\left(x^2 + 4x + \frac{7}{2}\right)$$

$$= 2\left(x^2 + 4x + \frac{8}{2} - \frac{1}{2}\right) = 2\left(x^2 + 4x + 4 - \frac{1}{2}\right) = \boxed{2(x+2)^2 - 1}$$

(c)

$$-2x^2 - 16x - 37 = -2\left(x^2 + 8x + \frac{37}{2}\right) = -2\left(x^2 + 8x + 16 + \frac{5}{2}\right)$$

$$= -2\left((x+4)^2 + \frac{5}{2}\right) = \boxed{-2(x+4)^2 - 5}$$

(d)

$$-2x^2 + 12x - 20 = -2\left(x^2 - 6x + 10\right) = -2\left(x^2 - 6x + 9 + 1\right)$$

$$= -2\left((x-3)^2 + 1\right) = \boxed{-2(x-3)^2 - 2}$$

(e)

$$0.5x^2 - 2x + 3 = 0.5\left(x^2 - 4x + 6\right) = 0.5\left(x^2 - 4x + 4 + 2\right)$$

$$= 0.5\left((x-2)^2 + 2\right) = \boxed{0.5(x-2)^2 + 1}$$

(f)

$$3x^2 - 3x - 3.25 = 3\left(x^2 - x - \frac{13}{12}\right) = 3\left(x^2 - x + \frac{1}{4} - \frac{4}{3}\right)$$
$$= 3\left((x - 0.5)^2 - \frac{4}{3}\right) = \boxed{3(x - 0.5)^2 - 4}$$

(g)

$$6x^2 + 36x + 53.5 = 6\left(x^2 + 6x + \frac{107}{12}\right) = 6\left(x^2 + 6x + 9 - \frac{1}{12}\right)$$
$$= 6\left((x + 3)^2 - \frac{1}{2}\right) = \boxed{6(x + 3)^2 - 0.5}$$

(h)

$$10x^2 + 20x + 10.25 = 10\left(x^2 + 2x + \frac{41}{40}\right) = 10\left(x^2 + 2x + 1 + \frac{1}{40}\right)$$
$$= 10\left((x + 1)^2 + 0.25\right) = \boxed{10(x + 1)^2 + 0.25}$$

(i)

$$2x^2 + 40x + 100 = 2\left(x^2 + 20x + 50\right) = 2\left(x^2 + 20x + 100 - 50\right)$$
$$= 2\left((x + 10)^2 - 50\right) = \boxed{2(x + 10)^2 - 100}$$

(j)

$$100x^2 - 2000x + 9900 = 100\left(x^2 - 20x + 99\right) = 100\left(x^2 - 20x + 100 - 1\right)$$
$$= 100\left((x - 10)^2 - 1\right) = \boxed{100(x - 10)^2 - 100}$$

Exercise 5.4

(a)
$$x^3 - 6x^2 + 12x - 9 = x^3 - 6x^2 + 12x - 8 - 1 = \boxed{(x - 2)^3 - 1}$$

(b)

$$2x^3 + 6x^2 + 6x + 4 = 2(x^3 + 3x^2 + 3x + 2)$$
$$= 2(x^3 + 3x^2 + 3x + 1 + 1) = \boxed{2(x + 1)^3 + 2}$$

(c)

$$-3x^3 - 18x^2 - 36x - 28 = -3\left(x^3 + 6x^2 + 12x + \frac{28}{3}\right)$$

$$= -3\left(x^3 + 6x^2 + 12x + 8 + \frac{4}{3}\right) = \boxed{-3(x+2)^3 - 4}$$

(d)

$$2x^3 - 18x^2 + 54x - 44 = 2\left(x^3 - 9x^2 + 27x - 22\right) = 2\left(x^3 - 9x^2 + 27x - 27 + 5\right)$$

$$= 2\left((x-3)^3 + 5\right) = \boxed{2(x-3)^3 + 10}$$

(e)

$$0.5x^3 - 1.5x^2 + 1.5x - 30.5 = 0.5\left(x^3 - 3x^2 + 3x - 61\right)$$

$$= 0.5\left(x^3 - 3x^2 + 3x - 1 - 60\right) = 0.5\left((x-1)^3 - 60\right) = \boxed{0.5(x-1)^3 - 30}$$

(f)

$$4x^3 + 48x^2 + 192x + 255.5 = 4\left(x^3 + 12x^2 + 48x + 63.875\right)$$

$$= 4\left(x^3 + 12x^2 + 48x + 64 - 0.125\right) = 4\left((x+4)^3 - 0.125\right) = \boxed{4(x+4)^3 - 0.5}$$

Exercise 5.5

(a)

$$\frac{x^2 + 2x}{x+1} = \frac{(x+1)^2 - 1}{x+1} = \boxed{x + 1 - \frac{1}{x+1}}$$

$$\begin{array}{r} x + 1 \\ x+1 \overline{\smash{)}\ x^2 + 2x} \\ \underline{-x^2 - x} \\ x \\ \underline{-x - 1} \\ -1 \end{array}$$

(b)

$$\frac{x^2 - 8x + 19}{x-4} = \frac{(x-4)^2 + 3}{x-4} = \boxed{x - 4 + \frac{3}{x-4}}$$

$$\begin{array}{r}
x - 4 \\
x-4 \overline{)\ x^2 - 8x + 19} \\
-x^2 + 4x \\
\hline
-4x + 19 \\
4x - 16 \\
\hline
3
\end{array}$$

(c)

$$\frac{-3x^2 + 30x - 71}{x^2 - 10x + 25} = \frac{-3(x-5)^2 + 4}{(x-5)^2} = \boxed{-3 + \frac{4}{(x-5)^2}}$$

$$\begin{array}{r}
-3 \\
x^2 - 10x + 25 \overline{)-3x^2 + 30x - 71} \\
3x^2 - 30x + 75 \\
\hline
4
\end{array}$$

(d)

$$\frac{3x^2 - 18x + 36}{3x - 9} = \frac{3(x-3)^2 + 9}{3(x-3)} = \boxed{x - 3 + \frac{3}{x - 3}}$$

$$\begin{array}{r}
x - 3 \\
3x - 9 \overline{)\ 3x^2 - 18x + 36} \\
-3x^2 + 9x \\
\hline
-9x + 36 \\
9x - 27 \\
\hline
9
\end{array}$$

(e)

$$\frac{2x^3 - 12x^2 + 24x - 14}{x^2 - 4x + 4} = \frac{2(x-2)^3 + 2}{(x-2)^2} = \boxed{2x - 4 + \frac{2}{(x-2)^2}}$$

$$\begin{array}{r}
2x - 4 \\
x^2 - 4x + 4 \overline{)\ 2x^3 - 12x^2 + 24x - 14} \\
-2x^3 + 8x^2 - 8x \\
\hline
-4x^2 + 16x - 14 \\
4x^2 - 16x + 16 \\
\hline
2
\end{array}$$

(f)

$$\frac{4x^3 + 36x^2 + 108x + 106}{2x + 6} = \frac{4(x+3)^3 - 2}{2x + 6} = \boxed{2x^2 + 12x + 18 - \frac{1}{x + 3}}$$

$$\begin{array}{r} 2x^2 \;\; + 12x \;\; + 18 \\ 2x+6\overline{\smash{\big)}\ 4x^3 + 36x^2 + 108x + 106} \\ -\,4x^3 - 12x^2 \\ \hline 24x^2 + 108x \\ -\,24x^2 \;\; -72x \\ \hline 36x + 106 \\ -\,36x - 108 \\ \hline -\,2 \end{array}$$

(g)

$$\frac{x^2 - 7}{x + 2} = \frac{(x-2)(x+2) - 3}{x + 2} = \boxed{x - 2 - \frac{3}{x + 2}}$$

$$\begin{array}{r} x - 2 \\ x+2\overline{\smash{\big)}\ x^2 -7} \\ -\,x^2 - 2x \\ \hline -\,2x - 7 \\ 2x + 4 \\ \hline -\,3 \end{array}$$

(h)

$$\frac{2x^2 - 9}{x - 2} = \frac{2(x-2)(x+2) - 1}{x - 2} = \boxed{2x + 4 - \frac{1}{x - 2}}$$

$$\begin{array}{r} 2x + 4 \\ x-2\overline{\smash{\big)}\ 2x^2 -9} \\ -\,2x^2 + 4x \\ \hline 4x - 9 \\ -\,4x + 8 \\ \hline -\,1 \end{array}$$

(i)

$$\frac{x^2 - x - 3}{x + 2} = \frac{(x+2)(x-3) + 3}{x + 2} = \boxed{x - 3 + \frac{3}{x + 2}}$$

$$\begin{array}{r} x - 3 \\ x + 2 \overline{)\ x^2\ - x - 3} \\ -x^2 - 2x \\ \hline -3x - 3 \\ 3x + 6 \\ \hline 3 \end{array}$$

(j)

$$\frac{2x^2 - 6x + 3}{x - 2} = \frac{2(x-1)(x-2) - 1}{x - 2} = \boxed{2x - 2 - \frac{1}{x - 2}}$$

$$\begin{array}{r} 2x - 2 \\ x - 2 \overline{)\ 2x^2 - 6x + 3} \\ -2x^2 + 4x \\ \hline -2x + 3 \\ 2x - 4 \\ \hline -1 \end{array}$$

Exercise 5.6

(a)

$$\frac{9x^2 - 12x - 11}{3x - 2} = \frac{(3x - 2)^2 - 15}{3x - 2} = \boxed{3x - 2 - \frac{15}{3x - 2}}$$

(b)

$$\frac{x^3 - 3x^2 + 3x - 1001}{x^2 - 2x + 1} = \frac{(x - 1)^3 - 1000}{(x - 1)^2} = \boxed{x - 1 - \frac{1000}{(x - 1)^2}}$$

(c)

$$\frac{8x^3 + 36x^2 + 54x + 27}{4x^2 + 12x + 9} = \frac{(2x + 3)^3}{(2x + 3)^2} = \boxed{3 + 2x}$$

(d)

$$\frac{x^3 - 6x^2y + 12xy^2 - 8y^3 - 1}{x - 2y - 1}$$
$$= \frac{(x - 2y)^3 - 1}{x - 2y - 1} = \frac{(x - 2y - 1)(x^2 - 4xy + x + 4y^2 - 2y + 1)}{x - 2y - 1}$$
$$= \boxed{x^2 + 4y^2 - 4xy + x - 2y + 1}$$

(e)

$$\frac{3x^6 + 3x^3 - 12x^2}{x^4 + x - 4} = \frac{3x^2(x^4 + x - 4)}{x^4 + x - 4} = \boxed{3x^2}$$

(f)

$$\frac{x^2 - 4x + 3}{x - 3} = \frac{(x-1)(x-3)}{x-3} = \boxed{x - 1}$$

(g)

$$\frac{x^2 - 3xy + 2y^2 + 1}{x - y} = \frac{(x-2y)(x-y)+1}{x-y} = \boxed{x - 2y + \frac{1}{x-y}}$$

(h)

$$\frac{x^2 - x - 13}{x - 4} = \frac{(x-4)(x+3)-1}{x-4} = \boxed{x + 3 - \frac{1}{x-4}}$$

(i)

$$\frac{2x^3 + 18x^2 + 54x + 57}{x + 3} = \frac{2(x+3)^3 + 3}{x+3} = \boxed{2x^2 + 12x + 18 + \frac{3}{x+3}}$$

(j)

$$\frac{4x^2 - y^2 + 1}{2x + y} = \frac{(2x+y)(2x-y)+1}{2x+y} = \boxed{2x - y + \frac{1}{2x+y}}$$

(k)

$$\frac{x^3 - 12x^2y + 48xy^2 - 64y^3}{x^2 - 8xy + 16y^2} = \frac{(x-4y)^3}{(x-4y)^2} = \boxed{x - 4y}$$

(l)

$$\frac{x^2 - 5xy^2 - 3xy + 15y^3}{x - 5y^2} = \frac{(x-3y)(x-5y^2)}{x-5y^2} = \boxed{x - 3y}$$

(m)

$$\frac{x^3 - 5x^2 - 2x - 10}{x^2 - 2} = \frac{(x^2-2)(x-5)-20}{x^2-2} = \boxed{x - 5 - \frac{20}{x^2-2}}$$

(n)

$$\frac{x^4 + 91}{x^2 - 3} = \frac{(x^2+3)(x^2-3)+100}{x^2-3} = \boxed{x^2 + 3 + \frac{100}{x^2-3}}$$

Exercise 5.7

(a)

$$
\begin{array}{r}
x^5 + 3x^4 \quad\ + x \\
x^3 + 2 \overline{)\ x^8 + 3x^7 + 2x^5 + 7x^4 + 2x} \\
\underline{-x^8 \qquad\quad - 2x^5 } \\
3x^7 \qquad\quad + 7x^4 \\
\underline{-3x^7 \qquad\quad - 6x^4 } \\
x^4 + 2x \\
\underline{-x^4 - 2x} \\
0
\end{array}
$$

(b)

$$
\require{enclose}
\begin{array}{r}
x^3 + x^2 + x + 1 \\
x-1 \enclose{longdiv}{\; x^4 \qquad\qquad\quad -6 } \\
\underline{-x^4 + x^3 \qquad\qquad} \\
x^3 \qquad\qquad \\
\underline{-x^3 + x^2 \qquad\quad} \\
x^2 \qquad \\
\underline{-x^2 + x \quad} \\
x - 6 \\
\underline{-x + 1} \\
-5
\end{array}
$$

(c)

$$
\begin{array}{r}
x^2 \;-x\; +3 \\
2x^3 - 3x^2 - 5 \enclose{longdiv}{\; 2x^5 - 5x^4 + 9x^3 - 14x^2 + 5x - 15 } \\
\underline{-2x^5 + 3x^4 \qquad\qquad +5x^2 \qquad\qquad} \\
-2x^4 + 9x^3 \;-9x^2 + 5x \qquad \\
\underline{2x^4 - 3x^3 \qquad\quad -5x \qquad} \\
6x^3 \;-9x^2 \qquad -15 \\
\underline{-6x^3 +9x^2 \qquad +15} \\
0
\end{array}
$$

(d)

$$
\begin{array}{r}
x^3 \qquad -2x + 1 \\
x^2 + 1 \enclose{longdiv}{\; x^5 \;-x^3 + x^2 \qquad -1 } \\
\underline{-x^5 \;-x^3 \qquad\qquad} \\
-2x^3 + x^2 \qquad \\
\underline{2x^3 \qquad +2x \quad} \\
x^2 + 2x - 1 \\
\underline{-x^2 \qquad -1} \\
2x - 2
\end{array}
$$

(e)

$$
\begin{array}{r}
3x^4 - 3x^3 + 2x^2 + x - 3 \\
x^3 + x^2 - 1 \overline{)\ 3x^7 \qquad - x^5 \qquad + x^3\ + x^2 \qquad + 1} \\
\underline{-3x^7 - 3x^6 \qquad + 3x^4} \\
-3x^6\ - x^5 + 3x^4\ + x^3 \\
\underline{3x^6 + 3x^5 \qquad - 3x^3} \\
2x^5 + 3x^4 - 2x^3\ + x^2 \\
\underline{-2x^5 - 2x^4 \qquad + 2x^2} \\
x^4 - 2x^3 + 3x^2 \\
\underline{-x^4\ - x^3 \qquad + x} \\
-3x^3 + 3x^2 + x + 1 \\
\underline{3x^3 + 3x^2 \qquad - 3} \\
6x^2 + x - 2
\end{array}
$$

(f)

$$
\begin{array}{r}
x^5 \qquad + x \\
x^4 + x^2 \overline{)\ x^9 + x^7 + x^5} \\
\underline{-x^9 - x^7} \\
x^5 \\
\underline{-x^5 - x^3} \\
-x^3
\end{array}
$$

(g)

$$
\begin{array}{r}
4x^3 - 11x^2\ + 56x - 226 \\
x + 4 \overline{)\ 4x^4\ + 5x^3 + 12x^2 \qquad - 2x \quad - 1} \\
\underline{-4x^4 - 16x^3} \\
-11x^3 + 12x^2 \\
\underline{11x^3 + 44x^2} \\
56x^2 \quad - 2x \\
\underline{-56x^2 - 224x} \\
-226x \quad - 1 \\
\underline{226x + 904} \\
903
\end{array}
$$

(h)

$$
\begin{array}{r}
2x^7 + 3x^6 - x^5 + 4x^4 \\
x^2 + x + 3 \overline{) \; 2x^9 + 5x^8 + 8x^7 + 12x^6 + x^5 + 12x^4} \\
-2x^9 - 2x^8 - 6x^7 \\
\hline
3x^8 + 2x^7 + 12x^6 \\
-3x^8 - 3x^7 - 9x^6 \\
\hline
-x^7 + 3x^6 + x^5 \\
x^7 + x^6 + 3x^5 \\
\hline
4x^6 + 4x^5 + 12x^4 \\
-4x^6 - 4x^5 - 12x^4 \\
\hline
0
\end{array}
$$

Exercise 5.8

(a)

$$(x+1) \circ (x-1) = (x-1) + 1 = \boxed{x}$$

(b)

$$(x^2+1) \circ (x+2) = (x+2)^2 + 1 = \boxed{x^2 + 4x + 5}$$

(c)

$$(3x^2+2) \circ (2x-1) = 3(2x-1)^2 + 2 = \boxed{12x^2 - 12x + 5}$$

(d)

$$(2x-1) \circ (3x^2+2) = 2(3x^2+2) - 1 = \boxed{6x^2 + 3}$$

(e)

$$(x-1) \circ (2x^3 - 3x^2) = 2x^3 - 3x^2 - 1 = \boxed{2x^3 - 3x^2 - 1}$$

(f)

$$(2x^3 - 3x^2) \circ (x-1) = 2(x-1)^3 - 3(x-1)^2 = \boxed{2x^3 - 9x^2 + 12x - 5}$$

(g)

$$(x^5)^{\circ 2} + (x+10)^{\circ 3} - 20 = (x^5) \circ (x^5) + (x+10) \circ (x+10) \circ (x+10) - 20$$

$$= \left((x^5)^5\right) + (((x+10)+10)+10) - 20 = \boxed{x^{25} + x + 10}$$

(h)

$$\sqrt[\circ]{x+20} + \sqrt[\circ]{x^{100}} = \sqrt[\circ]{(x+10)^{\circ 2}} + \sqrt[\circ]{(x^{10})^{\circ 2}} = \boxed{x^{10} + x + 10}$$

(i) We must find $(ax + b)$ such that

$$(ax + b) \circ (ax + b) = a(ax + b) + b = a^2x + ab + b$$

That is:

$$9x - 16 = a^2x + ab + b$$

This means that $a^2 = 9$. This is possible when $a = 3$ and $a = -3$. Hence, we have two options:

$$-3b + b = -2b = -16$$

and

$$3b + b = 4b = -16$$

We got two pairs: $a = -3$, $b = 8$ and $a = 3$, $b = -4$. Finally,

$$\sqrt[\circ]{9x - 16} = \sqrt[\circ]{(3x - 4)^{\circ 2}} = \boxed{3x - 4}$$

The other iterative root is $\boxed{-3x + 8}$.

(j)

$$\sqrt[\circ]{25x + 24} = \sqrt[\circ]{(5x + 4)^{\circ 2}} = \boxed{5x + 4}$$

The other iterative root is $\boxed{-5x - 6}$.

(k)

$$\sqrt[\circ]{16x + 15} = \sqrt[\circ]{(4x + 3)^{\circ 2}} = \boxed{4x + 3}$$

The other iterative root:

$$\sqrt[\circ]{16x + 15} = \boxed{-4x - 5}$$

(l)

$$\sqrt[\circ]{4x + 3} = \sqrt[\circ]{(2x + 1)^{\circ 2}} = \boxed{2x + 1}$$

The other iterative root:

$$\sqrt[\circ]{4x + 3} = \boxed{-2x - 3}$$

(m) This problem is easier to solve by recognizing the structure of the square $(x^2 - 1)^2 = x^4 - 2x^2 + 1$. That is $x^4 - 2x^2 = (x^2 - 1)^2 - 1$. This gives the way to the solution:

$$\sqrt[\circ]{x^4 - 2x^2} = \sqrt[\circ]{(x^2 - 1)^{\circ 2}} = \boxed{x^2 - 1}$$

(n)

$$\sqrt[\circ]{x^4 + 2x^2 + 2} = \sqrt[\circ]{(x^2 + 1)^{\circ 2}} = \boxed{x^2 + 1}$$

Exercise 5.9

(a)

$$\frac{(x^2+4)\circ(x+2)-4}{x+2} = \frac{((x+2)^2+4)-4}{x+2} = \boxed{x+2}$$

(b)

$$\frac{(x^3-x^2)\circ(x^2-1)}{(x^2-2)(x^2\circ(x^2-1))} = \frac{(x^2-1)^3-(x^2-1)^2}{(x^2-2)(x^2-1)^2} = \frac{(x^2-1)^2(x^2-1-1)}{(x^2-2)(x^2-1)^2} = \boxed{1}$$

(c)

$$\frac{(x+1)^2\circ(x-2)^2}{x^2-4x+5} - x(x-1)^{\circ 4} = \frac{((x-2)^2+1)^2}{x^2-4x+5} - x((((x-1)-1)-1)-1) = \boxed{5}$$

(d)

$$\frac{2x^3\circ(x+1)\circ x^2}{x^4+2x^2+1} - (2x)\circ x^2 = \frac{2x^3\circ(x^2+1)}{x^4+2x^2+1} - 2x^2 = \frac{2(x^2+1)^3}{x^4+2x^2+1} - 2x^2$$

$$= \frac{2x^6+6x^4+6x^2+2}{x^4+2x^2+1} - 2x^2 = 2x^2+2-2x^2 = \boxed{2}$$

(e)

$$\frac{x^2\circ x^3\circ(x+1)}{x^5\circ(x+1)} = \frac{x^2\circ(x+1)^3}{(x+1)^5} = \frac{((x+1)^3)^2}{(x+1)^5} = \boxed{x+1}$$

(f)

$$\frac{(x+2)^2\circ(x-3)}{(x-3)\circ(x+2)^2}\cdot\frac{(x^2-3)\circ(x+2)}{(x-1)\circ x} = \frac{(x-3)+2)^2}{(x+2)^2-3}\cdot\frac{(x+2)^2-3}{x-1} = \boxed{x-1}$$

(g)

$$(x^2-2)^{\circ 2} - (x^2+2)^{\circ 2} + 8\left(x^{\circ 2}\right)^2$$

$$= (x^2-2)\circ(x^2-2) - (x^2+2)\circ(x^2+2) + 8(x\circ x)^2$$

$$= ((x^2-2)^2-2) - ((x^2+2)^2+2) + 8x^2 = \boxed{-4}$$

(h)

$$\frac{(x-2)(x^2-4)^{\circ 2}}{(x^2-6)(x-1)^{\circ 2}} = \frac{(x-2)((x^2-4)^2-4)}{(x^2-6)((x-1)-1)} = \boxed{x^2-2}$$

(i)

$$\left(x^2\circ(x+2)^{\circ 3}\circ(x+1)^{\circ 3}\right)\left(\sqrt[\circ]{x+18}\right)^{-1}$$

$$= \left(x^2\circ(x+6)\circ(x+3)\right)\left(\sqrt[\circ]{(x+9)^{\circ 2}}\right)^{-1}$$

$$= (x^2\circ(x+9))(x+9)^{-1} = (x+9)^2(x+9)^{-1} = \boxed{x+9}$$

(j)

$$\sqrt[4]{\sqrt[\circ 3]{(x^8)^{\circ 2}}} = \sqrt[4]{\sqrt[\circ 3]{(x^8)^8}} = \sqrt[4]{\sqrt[\circ 3]{x^{64}}} = \sqrt[4]{\sqrt[\circ 3]{((x^4)^4)^4}} = \sqrt[4]{\sqrt[\circ 3]{(x^4)^{\circ 3}}} = \sqrt[4]{x^4} = \boxed{|x|}$$

Exercise 5.10

(a)

$$\frac{(x^3 + x^2)^{\circ 2}}{(x+1)^2(x^3 + x^2 + 1)} = \frac{(x^3 + x^2)^3 + (x^3 + x^2)^2}{(x+1)^2(x^3 + x^2 + 1)} = \frac{(x^3 + x^2)^2(x^3 + x^2 + 1)}{(x+1)^2(x^3 + x^2 + 1)}$$

$$= \frac{x^4(x+1)^2}{(x+1)^2} = \boxed{x^4}$$

(b)

$$\frac{(2x^3 - 1) \circ (2x^2 + 1) - 12x^2 - 1}{2x^2 + 3} = \frac{2(2x^2 + 1)^3 - 1 - 12x^2 - 1}{2x^2 + 3} = \frac{16x^6 + 24x^4}{2x^2 + 3}$$

$$= \frac{8x^4(2x^2 + 3)}{2x^2 + 3} = \boxed{8x^4}$$

Exercise 5.11

$$\frac{x^2((x^2 - 2)x^2 + 1) - 3}{(x-1)^2(x+1)^2} - \frac{(x^3 - 3) \circ (x^2 - 1)}{x^4 - 2x^2 + 1}$$

$$= \frac{x^6 - 2x^4 + x^2 - 3}{(x^2 - 1)^2} - \frac{(x^2 - 1)^3 - 3}{(x^2 - 1)^2}$$

$$= \frac{x^6 - 2x^4 + x^2 - 3 - (x^6 - 3x^4 + 3x^2 - 4)}{(x^2 - 1)^2}$$

$$= \frac{x^4 - 2x^2 + 1}{(x^2 - 1)^2} = \frac{(x^2 - 1)^2}{(x^2 - 1)^2} = \boxed{1}$$

Exercise 5.12

$$\frac{(x+1) \circ (x+1)^4}{(x^2 + 2x + 2) \circ (x^2 + 2x)} = \frac{(x+1)^4 + 1}{(x^2 + 2x)^2 + 2(x^2 + 2x) + 2}$$

$$= \frac{(x^2 + 2x + 1)^2 + 1}{(x^2 + 2x)^2 + 2(x^2 + 2x) + 2} = \frac{(x^2 + 2x)^2 + 2(x^2 + 2x) + 1 + 1}{(x^2 + 2x)^2 + 2(x^2 + 2x) + 2} = \boxed{1}$$

Exercise 5.13

$$A = x^4 \circ (2x - 1) - x = (2x - 1)^4 - x = 16x^4 - 32x^3 + 24x^2 - 9x + 1$$

$$B \circ C = 16\left(x - \frac{1}{3}\right)^3 + \frac{8}{3}\left(x - \frac{1}{3}\right) + \frac{13}{27} = 16x^3 - 16x^2 + 8x - 1$$

$$
\begin{array}{r}
x - 1 \\
16x^3 - 16x^2 + 8x - 1 \overline{\smash{)}\ 16x^4 - 32x^3 + 24x^2 - 9x + 1} \\
-16x^4 + 16x^3 - 8x^2 + x \\
\hline
-16x^3 + 16x^2 - 8x + 1 \\
16x^3 - 16x^2 + 8x - 1 \\
\hline
0
\end{array}
$$

Thus:

$$\frac{A}{B \circ C} = \boxed{x - 1}$$

Exercise 5.14

$$A = (x^3 - x^2)^3 - (x^3 - x^2)^2 = (x^3 - x^2)^2(x^3 - x^2 - 1)$$

$$\frac{A}{B} = \frac{(x^3 - x^2)^2(x^3 - x^2 - 1)}{x^6 - 2x^5 + x^4} = \frac{(x^3 - x^2)^2(x^3 - x^2 - 1)}{(x^3 - x^2)^2} = x^3 - x^2 - 1$$

We find C using long division:

$$
\begin{array}{r}
x^3 - x^2 \\
x^2 + x + 1 \overline{\smash{)}\ x^5 \qquad\quad - x^2} \\
-x^5 - x^4 - x^3 \\
\hline
-x^4 - x^3 - x^2 \\
x^4 + x^3 + x^2 \\
\hline
0
\end{array}
$$

Thus:

$$\frac{A}{B} - C = x^3 - x^2 - 1 - (x^3 - x^2) = \boxed{-1}$$

Exercise 5.15

$$A = (x^2 + 1)^2 + 1 = x^4 + 2x^2 + 2$$

$$A^{\circ 2} = (x^4 + 2x^2 + 2)^4 + 2(x^4 + 2x^2 + 2)^2 + 2$$

$$C \circ D = (x^4 + 2x^2)^2 + 4(x^4 + 2x^2)$$

$$B \circ C \circ D = ((x^4 + 2x^2)^2 + 4(x^4 + 2x^2))^2 + 10((x^4 + 2x^2)^2 + 4(x^4 + 2x^2)) + 26$$

Denote $a = x^4 + 2x^2$:

$$\frac{A^{\circ 2}}{B \circ C \circ D} = \frac{(a + 2)^4 + 2(a + 2)^2 + 2}{(a^2 + 4a)^2 + 10(a^2 + 4a) + 26}$$

Notice that $a^2 + 4a = (a + 2)^2 - 4$ and denote $b = (a + 2)^2$:

$$\frac{b^2 + 2b + 2}{(b - 4)^2 + 10(b - 4) + 26} = \frac{b^2 + 2b + 2}{b^2 + 2b + 2} = \boxed{1}$$

Exercise 5.16

(a) Consider:
$$(x + 1) \circ (x + 2) = x + 3$$
$$(x + 1) \circ x + (x + 1) \circ 2 = x + 1 + 2 + 1 = x + 4$$

(b) Consider:
$$x \circ (x + 2) = x + 2$$
$$x \circ x + x \circ 2 = x + 2$$

(c) Consider:
$$(x + 1) \circ (x + 2) = x + 3$$
$$(x + 2) \circ (x + 1) = x + 3$$

Exercise 5.17

The expression $(x + 1) \circ (x + 2)$ is equal to $(x + 2) + 1$. Therefore, if we replace $x + 2$ with k, we must obtain: $(x + 1) \circ (x + 2) = k + 1$.

The correct transition to the new variable k would be:

$$(x + 1) \circ (x + 2) = (x + 1) \circ (x + 2) \circ (k - 2) = (x + 1) \circ k = k + 1$$

In other words, the act of replacing x with $k - 2$ is a composition. An error committed in the proposed solution was applying this composition twice – first to the first polynomial and then to the second. For a composition of three polynomials:

$$A \circ B \circ C \neq A \circ C \circ B \circ C$$

However, the proposed solution did the following:

$$(x + 1) \circ (x + 2) \circ (k - 2) = [(x + 1) \circ (k - 2)] \circ [(x + 2) \circ (k - 2)]$$

Exercise 5.18

(a)
$$(x + 1)^{\circ 2} = (x + 1) \circ (x + 1) = x + 2$$
$$(x + 1)^{\circ 3} = (x + 1) \circ (x + 2) = x + 3$$
$$(x + 1)^{\circ 4} = (x + 1) \circ (x + 3) = x + 4$$
$$(x + 1)^{\circ 5} = (x + 1) \circ (x + 4) = x + 5$$

We notice that
$$\boxed{(x + 1)^{\circ n} = x + n}$$

(b) Let's iterate to see the pattern:

$$(2x + 1)^{\circ 2} = (2x + 1) \circ (2x + 1) = 4x + 3$$

$$(2x + 1)^{\circ 3} = (2x + 1) \circ (4x + 3) = 8x + 7$$

$$(2x + 1)^{\circ 4} = (2x + 1) \circ (8x + 7) = 16x + 15$$

$$(2x + 1)^{\circ 5} = (2x + 1) \circ (16x + 15) = 32x + 31$$

We can notice that
$$\boxed{(2x + 1)^{\circ n} = 2^n x + 2^n - 1}$$

(c)
$$\sqrt[\circ 3]{x + 45} = \sqrt[\circ 3]{(x + 15)^{\circ 3}} = \boxed{x + 15}$$

(d)
$$\sqrt[\circ 3]{27x + 13} = \sqrt[\circ 3]{(3x + 1)^{\circ 3}} = \boxed{3x + 1}$$

Exercise 5.19

(a) If the variable x in a one-variable polynomial is replaced by x, the polynomial does not change:

$$Polynomial \circ x = Polynomial$$

And also:

$$x \circ Polynomial = Polynomial$$

Thus, $\boxed{E = x}$.

(b) We must find $ax + b$ that satisfies $(ax + b) \circ (2x + 1) = x$. Since

$$(ax + b) \circ (2x + 1) = 2ax + a + b,$$

it must be that:

$$2ax + a + b = x$$

That is $a = 0.5$, and $b = -0.5$. Hence:

$$(2x + 1)^{\circ -1} = \boxed{0.5x - 0.5}$$

Let's check that $A^{\circ -1} \circ A = A \circ A^{\circ -1} = x$:

$$(2x + 1) \circ (0.5x - 0.5) = 2(0.5x - 0.5) + 1 = x$$

$$(0.5x - 0.5) \circ (2x + 1) = 0.5(2x + 1) - 0.5 = x$$

(c)

$$(8x + 7)^{\circ \frac{2}{3}} = \left(\sqrt[\circ 3]{8x + 7} \right)^{\circ 2} = (2x + 1)^{\circ 2} = 4x + 3$$

$$(27x + 13)^{\circ -\frac{1}{3}} = \left(\sqrt[\circ 3]{27x + 13} \right)^{\circ -1} = (3x + 1)^{\circ -1} = \frac{1}{3}x - \frac{1}{3}$$

$$4x + 3 + \frac{1}{3}x - \frac{1}{3} = \boxed{\frac{13}{3}x + \frac{8}{3}}$$

Exercise 5.20

(a) Long division for $n = 2$ (can be easily divided without long division):

$$
\begin{array}{r}
x + 1 \\
x - 1 \overline{\smash{\big)}\ x^2 \quad\quad - 1} \\
\underline{-\ x^2 + x} \\
x - 1 \\
\underline{-\ x + 1} \\
0
\end{array}
$$

Long division for $n = 3$ (can be easily divided without long division):

$$
\begin{array}{r}
x^2 + x + 1 \\
x - 1 \overline{\smash{\big)}\ x^3 \quad\quad\quad\quad - 1} \\
\underline{-\ x^3 + x^2} \\
x^2 \\
\underline{-\ x^2 + x} \\
x - 1 \\
\underline{-\ x + 1} \\
0
\end{array}
$$

Long division for $n = 4$ (can be easily divided without long division):

$$
\begin{array}{r}
x^3 + x^2 + x + 1 \\
x - 1 \overline{\smash{\big)}\ x^4 \quad\quad\quad\quad\quad\quad - 1} \\
\underline{-\ x^4 + x^3} \\
x^3 \\
\underline{-\ x^3 + x^2} \\
x^2 \\
\underline{-\ x^2 + x} \\
x - 1 \\
\underline{-\ x + 1} \\
0
\end{array}
$$

Long division for $n = 5$:

$$
\begin{array}{r}
x^4 + x^3 + x^2 + x + 1 \\
x - 1 \overline{\smash{\big)}\ x^5 \qquad\qquad\qquad\quad\ -1} \\
\underline{-\ x^5 + x^4} \\
x^4 \\
\underline{-\ x^4 + x^3} \\
x^3 \\
\underline{-\ x^3 + x^2} \\
x^2 \\
\underline{-\ x^2 + x} \\
x - 1 \\
\underline{-\ x + 1} \\
0
\end{array}
$$

Thus:

$$\frac{x - 1}{x - 1} = 1$$

$$\frac{x^2 - 1}{x - 1} = x + 1$$

$$\frac{x^3 - 1}{x - 1} = x^2 + x + 1$$

$$\frac{x^4 - 1}{x - 1} = x^3 + x^2 + x + 1$$

$$\frac{x^5 - 1}{x - 1} = x^4 + x^3 + x^2 + x + 1$$

We see that for any n:

$$\boxed{\ \frac{x^n - 1}{x - 1} = x^{n-1} + x^{n-2} + x^{n-3} + \cdots + x + 1\ }$$

What we have done is not a "a proof" of this formula of course: we simply inferred the general structure from a few iterations. The proof is not difficult though:

$$
\begin{aligned}
\frac{x^n - 1}{x - 1} &= \frac{x^n - x^{n-1} + x^{n-1} - x^{n-2} + x^{n-2} - \cdots + x^2 - x + x - 1}{x - 1} \\
&= \frac{x^n - x^{n-1}}{x - 1} + \frac{x^{n-1} - x^{n-2}}{x - 1} + \cdots + \frac{x^2 - x}{x - 1} + \frac{x - 1}{x - 1} \\
&= x^{n-1} + x^{n-2} + x^{n-3} + \cdots + x + 1
\end{aligned}
$$

(b)

Long division for $n = 2$:

$$
\require{enclose}
\begin{array}{r}
x - 1 \\
x+1 \enclose{longdiv}{\; x^2 + 1} \\
\underline{-\,x^2 - x } \\
-\,x + 1 \\
\underline{x + 1} \\
2
\end{array}
$$

Long division for $n = 3$:

$$
\require{enclose}
\begin{array}{r}
x^2 - x + 1 \\
x+1 \enclose{longdiv}{\; x^3 + 1} \\
\underline{-\,x^3 - x^2 } \\
-\,x^2 \\
\underline{x^2 + x } \\
x + 1 \\
\underline{-\,x - 1} \\
0
\end{array}
$$

Long division for $n = 4$:

$$
\require{enclose}
\begin{array}{r}
x^3 - x^2 + x - 1 \\
x+1 \enclose{longdiv}{\; x^4 + 1} \\
\underline{-\,x^4 - x^3 } \\
-\,x^3 \\
\underline{x^3 + x^2 } \\
x^2 \\
\underline{-\,x^2 - x } \\
-\,x + 1 \\
\underline{x + 1} \\
2
\end{array}
$$

Long division for $n = 5$:

$$
\begin{array}{r}
x^4 - x^3 + x^2 - x + 1 \\
x + 1 \overline{\smash{)}\ x^5 \qquad\qquad\qquad +1} \\
\underline{-\,x^5 - x^4} \\
-\,x^4 \\
\underline{x^4 + x^3} \\
x^3 \\
\underline{-\,x^3 - x^2} \\
-\,x^2 \\
\underline{x^2 + x} \\
x + 1 \\
\underline{-\,x - 1} \\
0
\end{array}
$$

Long division for $n = 6$:

$$
\begin{array}{r}
x^5 - x^4 + x^3 - x^2 + x - 1 \\
x + 1 \overline{\smash{)}\ x^6 \qquad\qquad\qquad\qquad +1} \\
\underline{-\,x^6 - x^5} \\
-\,x^5 \\
\underline{x^5 + x^4} \\
x^4 \\
\underline{-\,x^4 - x^3} \\
-\,x^3 \\
\underline{x^3 + x^2} \\
x^2 \\
\underline{-\,x^2 - x} \\
-\,x + 1 \\
\underline{x + 1} \\
2
\end{array}
$$

Thus:

$$\frac{x+1}{x+1} = 1$$

$$\frac{x^2+1}{x+1} = x - 1 + \frac{2}{x+1}$$

$$\frac{x^3+1}{x+1} = x^2 - x + 1$$

$$\frac{x^4+1}{x+1} = x^3 - x^2 + x - 1 + \frac{2}{x+1}$$

$$\frac{x^5+1}{x+1} = x^4 - x^3 + x^2 - x + 1$$

$$\frac{x^6+1}{x+1} = x^5 - x^4 + x^3 - x^2 + x - 1 + \frac{2}{x+1}$$

We see that for odd numbers n:

$$\frac{x^n+1}{x+1} = x^{n-1} - x^{n-2} + x^{n-3} - \cdots - x + 1$$

This can also be written as

$$\boxed{\frac{x^n+1}{x+1} = (-x)^{n-1} + (-x)^{n-2} + (-x)^{n-3} + \cdots + (-x)^1 + 1}$$

For even numbers n:

$$\boxed{\frac{x^n+1}{x+1} = (-1)^n x^{n-1} + (-1)^{n-1} x^{n-2} + (-1)^{n-2} x^{n-3} + \cdots + (-1)^2 x^1 - 1 + \frac{2}{x+1}}$$

Solutions for Chapter 6 Exercises

Exercise 6.1

(a)

$$x_1^2 + x_2^2 + \cdots + x_n^2 = \boxed{\sum_{i=1}^{n} x_i^2}$$

(b)

$$x_1 y_2 + x_2 y_3 + \cdots + x_{n-1} y_n = \boxed{\sum_{i=1}^{n-1} x_i y_{i+1}} = \boxed{\sum_{i=2}^{n} x_{i-1} y_i}$$

(c)

$$4\sqrt{A_5} + 5\sqrt{A_6} + \cdots + 99\sqrt{A_{100}} = \boxed{\sum_{i=5}^{100} (i-1)\sqrt{A_i}} = \boxed{\sum_{i=4}^{99} i\sqrt{A_{i+1}}}$$

(d)

$$(1.02\gamma_5)(1.03\gamma_6)\cdots(1.77\gamma_{80}) = \boxed{\prod_{i=5}^{80} \left(1 + \frac{i-3}{100}\right)\gamma_i} = \boxed{\prod_{i=2}^{77} \left(1 + \frac{i}{100}\right)\gamma_{i+3}}$$

(e)

$$\frac{z^2}{3} + \frac{z^3}{4} + \cdots + \frac{z^{23}}{24} = \boxed{\sum_{i=2}^{23} \frac{z^i}{i+1}} = \boxed{\sum_{i=3}^{24} \frac{z^{i-1}}{i}}$$

(f)

$$\frac{-2-1+1+2+\cdots+k}{(k-1)^2(k-2)^3\cdots 1} = \boxed{\frac{\sum_{i=-2}^{k} i}{\prod_{i=1}^{k-1}(k-i)^{i+1}}}$$

(g)

$$-2+4-6+8-\cdots+200 = \boxed{2\sum_{i=1}^{100}(-1)^i i}$$

(h)

$$(-2)(3)(-4)(5)\cdots(99) = \boxed{\prod_{i=2}^{99}(-1)^{i+1}i}$$

(i)

$$729 - 512 + 343 - \cdots - 343 + 512 - 729$$

$$= 9^3 - 8^3 + 7^3 - \cdots - 7^3 + 8^3 - 9^3 = \boxed{\sum_{i=-9}^{9}(-1)^i i^3}$$

Notice that this sum is equal to $\boxed{0}$.

(j)

$$(-0.5)(1.5)(-2.5)(3.5)\cdots(-30.5)$$

$$= \left(-\frac{2\cdot 0+1}{2}\right)\left(\frac{2\cdot 1+1}{2}\right)\left(-\frac{2\cdot 2+1}{2}\right)\left(\frac{2\cdot 3+1}{2}\right)\cdots\left(-\frac{2\cdot 30+1}{2}\right)$$

$$= \boxed{\prod_{i=0}^{30}(-1)^{i+1}\frac{2i+1}{2}}$$

(k)

$$a_1^{1+2+\cdots+n} + a_2^{1+2+\cdots+n} + \cdots + a_m^{1+2+\cdots+n}$$

$$= (a_1 a_1^2 \cdots a_1^n) + (a_2 a_2^2 \cdots a_2^n) + \cdots + (a_m a_m^2 \cdots a_m^n)$$

$$= \prod_{j=1}^{n}a_1^j + \prod_{j=1}^{n}a_2^j + \cdots + \prod_{j=1}^{n}a_m^j = \boxed{\sum_{i=1}^{m}\prod_{j=1}^{n}a_i^j}$$

But also:

$$a_1^{1+2+\cdots+n} + a_2^{1+2+\cdots+n} + \cdots + a_m^{1+2+\cdots+n}$$

$$= a_1^{n(n+1)/2} + a_2^{n(n+1)/2} + \cdots + a_m^{n(n+1)/2} = \boxed{\sum_{i=1}^{m}a_i^{n(n+1)/2}}$$

(l)

$$1 \cdot 3 \cdot 6 \cdot 10 \cdots 20100 = 1 \cdot 3 \cdot 6 \cdot 10 \cdots \frac{200 \cdot 201}{2}$$

$$= (1) \cdot (1+2) \cdot (1+2+3) \cdot (1+2+3+4) \cdots (1+2+3+4+\cdots+200)$$

$$= \left(\sum_{j=1}^{1} j \right) \left(\sum_{j=1}^{2} j \right) \left(\sum_{j=1}^{3} j \right) \cdots \left(\sum_{j=1}^{200} j \right) = \boxed{\prod_{i=1}^{200} \sum_{j=1}^{i} j} = \boxed{\frac{1}{2} \prod_{i=1}^{200} i(i+1)}$$

Exercise 6.2

(a)

$$\sum_{k=1}^{4} \frac{W_k}{k^2 + 1} = \frac{W_1}{1^2 + 1} + \frac{W_2}{2^2 + 1} + \frac{W_3}{3^2 + 1} + \frac{W_4}{4^2 + 1}$$

$$= \frac{W_1}{2} + \frac{W_2}{5} + \frac{W_3}{10} + \frac{W_4}{17} = \frac{4}{2} - \frac{5}{5} + \frac{50}{10} - \frac{34}{17} = \boxed{4}$$

(b)

$$\prod_{s=1}^{4} (-1)^{s+1} \sqrt{s + r_s} = \left(\sqrt{1 + r_1} \right) \left(-\sqrt{2 + r_2} \right) \left(\sqrt{3 + r_3} \right) \left(-\sqrt{4 + r_4} \right)$$

$$= \left(\sqrt{1 + 3} \right) \left(-\sqrt{2 - 1} \right) \left(\sqrt{3 + 6} \right) \left(-\sqrt{4 + 12} \right) = \boxed{24}$$

(c)

$$\sum_{i=1}^{5} (-1)^i x_i^{i-3} = -x_1^{-2} + x_2^{-1} - x_3^0 + x_4^1 - x_5^2$$

$$= -0.25^{-2} + (-0.5)^{-1} - 987^0 + (-3)^1 - 10^2 = -16 - 2 - 1 - 3 - 100 = \boxed{-122}$$

(d)

$$\prod_{j=1}^{5} (-1)^j (j\phi_j)^{j-5} = \left(-(\phi_1)^{-4} \right) \left((2\phi_2)^{-3} \right) \left(-(3\phi_3)^{-2} \right) \left((4\phi_4)^{-1} \right) \left(-(5\phi_5)^0 \right)$$

$$= \left(-\frac{1}{\phi_1^4} \right) \left(\frac{1}{(2\phi_2)^3} \right) \left(-\frac{1}{(3\phi_3)^2} \right) \left(\frac{1}{4\phi_4} \right) (-1)$$

$$= \left(-\frac{1}{(1/2)^4} \right) \left(\frac{1}{(2/6)^3} \right) \left(-\frac{1}{(3/12)^2} \right) \left(\frac{1}{4 \cdot 1728} \right) (-1)$$

$$= -\frac{(-2^4)3^3(-4^2)}{4 \cdot 1728} = \boxed{-1}$$

(e)

$$\sum_{r=1}^{4} r^r - \sum_{r=-2}^{3} (-1)^r r^3 = (1^1 + 2^2 + 3^3 + 4^4) - ((-2)^3 - (-1)^3 + 0^3 - 1^3 + 2^3 - 3^3) = \boxed{315}$$

(f)

$$2^{-4} \left(\prod_{j=-4}^{-2} (2j + 2) - \prod_{j=2}^{4} 2 \left(j + (-1)^{j+1} \right)^{j-1} \right)$$

$$= 2^{-4} \left(2^3 \prod_{j=-4}^{-2} (j + 1) - 2^3 \prod_{j=2}^{4} (j + (-1)^{j+1})^{j-1} \right)$$

$$= \frac{(-4 + 1)(-3 + 1)(-2 + 1) - (2 - 1)^1 (3 + 1)^2 (4 - 1)^3}{2} = \boxed{-219}$$

(g) Using Formulas 6.9 and 6.10:

$$\sum_{k=25}^{50} k = \sum_{k=1}^{50} k - \sum_{k=1}^{24} k = \frac{50(50 + 1)}{2} - \frac{24(24 + 1)}{2} = 975$$

$$\sum_{k=10}^{15} k^2 = \sum_{k=1}^{15} k^2 - \sum_{k=1}^{9} k^2 = \frac{15(15 + 1)(2 \cdot 15 + 1)}{6} - \frac{9(9 + 1)(2 \cdot 9 + 1)}{6} = 955$$

$$\sum_{k=25}^{50} k - \sum_{k=10}^{15} k^2 = 975 - 955 = \boxed{20}$$

Exercise 6.3

(a)

$$\sum_{j=1}^{4} \sum_{i=1}^{j} (-1)^j i^2 = \sum_{i=1}^{1} (-1)^1 i^2 + \sum_{i=1}^{2} (-1)^2 i^2 + \sum_{i=1}^{3} (-1)^3 i^2 + \sum_{i=1}^{4} (-1)^4 i^2$$

$$= -1 + \sum_{i=1}^{2} i^2 - \sum_{i=1}^{3} i^2 + \sum_{i=1}^{4} i^2 = \boxed{20}$$

(b)

$$\sum_{k=-2}^{2} \prod_{s=0}^{2} (k + (-1)^{ks} s) = \sum_{k=-2}^{2} k(k + (-1)^k)(k + 2) = \boxed{26}$$

(c)

$$\prod_{j=1}^{2}\prod_{i=0}^{j}(1+ij)(-1)^{1+ij}$$

$$= \left(\prod_{i=0}^{1}(1+i)(-1)^{1+i}\right)\left(\prod_{i=0}^{2}(1+2i)(-1)^{1+2i}\right)$$

$$= -\left(\prod_{i=0}^{1}(1+i)(-1)^{1+i}\right)\left(\prod_{i=0}^{2}(1+2i)\right) = -(-2)15 = \boxed{30}$$

(d)

$$6^{-3}\prod_{k=-3}^{-1}\sum_{s=-2}^{0}2ks = 6^{-3}2^3\prod_{k=-3}^{-1}\sum_{s=-2}^{0}ks = 6^{-3}2^3\prod_{k=-3}^{-1}(-3k) = -6^{-3}2^3 3^3\prod_{k=-3}^{-1}k = \boxed{6}$$

Exercise 6.4

(a)

$$\frac{n!}{(n-1)!} + \frac{(2n)!!}{(2n-2)!!} = \frac{1\cdot 2\cdot 3\cdots(n-1)\cdot n}{1\cdot 2\cdot 3\cdots(n-1)} + \frac{2\cdot 4\cdot 6\cdots(2n-2)\cdot(2n)}{2\cdot 4\cdot 6\cdots(2n-2)} = \boxed{3n}$$

(b)

$$\frac{(2n+1)!!}{(2n-1)!!} - \frac{(2n)!}{(2n-1)!} = \frac{1\cdot 3\cdot 5\cdots(2n-1)\cdot(2n+1)}{1\cdot 3\cdot 5\cdots(2n-1)} - \frac{1\cdot 2\cdot 3\cdots(2n-1)\cdot(2n)}{1\cdot 3\cdot 5\cdots(2n-1)} = \boxed{1}$$

(c)

$$\frac{(4!+6)n!28!!}{(n-1)!30!!} = \frac{(1\cdot 2\cdot 3\cdot 4+6)(1\cdot 2\cdot 3\cdots(n-1)\cdot n)(2\cdot 4\cdot 6\cdots 28)}{(1\cdot 2\cdot 3\cdots(n-1))(2\cdot 4\cdot 6\cdots 28\cdot 30)} = \frac{30n}{30} = \boxed{n}$$

(d)

$$\frac{87!!(n+2)!}{(4!+5!!+6!!)(n+1)!85!!}$$

$$= \frac{(3\cdot 5\cdots 85\cdot 87)(2\cdot 3\cdots(n+1)\cdot(n+2))}{(2\cdot 3\cdot 4+3\cdot 5+2\cdot 4\cdot 6)(2\cdot 3\cdots(n+1))(3\cdot 5\cdots 85)}$$

$$= \frac{87(n+2)}{24+15+48} = \boxed{n+2}$$

Exercise 6.5

(a)

$$\sum_{n=1}^{k} \frac{(2n)!!(2n+1)!!}{(k+2)(2n)!} = \frac{1}{k+2} \sum_{n=1}^{k} \frac{(2n+1)!}{(2n)!}$$

$$= \frac{1}{k+2} \sum_{n=1}^{k} (2n+1) = \frac{1}{k+2} \left(\sum_{n=1}^{k} 2n + \sum_{n=1}^{k} 1 \right)$$

$$= \frac{2}{k+2} \sum_{n=1}^{k} n + \frac{1}{k+2} \sum_{n=1}^{k} 1 = \frac{2}{k+2} \cdot \frac{k(k+1)}{2} + \frac{k}{k+2} = \boxed{k}$$

(b)

$$\frac{1}{(s-1)!} \left(\sum_{x=s-1}^{s} \frac{(x!)!}{(x!-1)!} \right) = \frac{1}{(s-1)!} \left(\sum_{x=s-1}^{s} x! \right) = \frac{(s-1)! + s!}{(s-1)!} = \boxed{s+1}$$

(c)

$$\sum_{d=0}^{2} \left(\frac{(2d^2)!!}{2(2d^2-2)!!} \right)! = \sum_{d=0}^{2} (d^2)! = 0! + 1! + 4! = 1 + 1 + 24 = \boxed{26}$$

(d)

$$\left(\frac{1}{2} \prod_{n=1}^{2} \frac{(n^2+1)!}{2(n^2-1)!} \right)!! = \left(\frac{1}{2 \cdot 2^2} \prod_{n=1}^{2} n^2(n^2+1) \right)!! = \left(\frac{40}{2 \cdot 2^2} \right)!! = 5!! = \boxed{15}$$

(e)

$$\frac{(5!!)!!}{17!!} \sum_{m=0}^{2} \left(xm! + \sum_{n=0}^{2} \frac{x(mn)!}{m!n!} \right) = \frac{1}{17} \left(x \sum_{m=0}^{2} m! + x \sum_{m=0}^{2} \sum_{n=0}^{2} \frac{(mn)!}{m!n!} \right)$$

$$= \frac{1}{17} \left(4x + x \sum_{m=0}^{2} \frac{2m! + (2m)! + 2}{2m!} \right) = \frac{1}{17} (4x + 13x) = \boxed{x}$$

Exercise 6.6

$$\frac{1}{k!} \prod_{n=1}^{k} \frac{1}{(n-1)!} \sqrt[4]{\frac{((n!)^2 + n!)^3 - (n! - (n!)^2)^3}{2((n!)^2 + 3)}}$$

$$= \frac{1}{k!} \prod_{n=1}^{k} \frac{1}{(n-1)!} \sqrt[4]{(n!)^4} = \frac{1}{k!} \prod_{n=1}^{k} \frac{n!}{(n-1)!} = \frac{1}{k!} \prod_{n=1}^{k} n = \frac{k!}{k!} = \boxed{1}$$

Exercise 6.7

$$A_k = \left(\sqrt{2k^{-1}} - \sqrt{k}\right)^2 \left(\sqrt{2} + k\right)^2 - k^3 = \frac{4(1 - k^2)}{k}$$

$$B_k = \sum_{i=1}^{k}(2i - 1) = 2\sum_{i=1}^{k} i - \sum_{i=1}^{k} 1 = k(k + 1) - k = k^2$$

Then:

$$A_k B_k = \frac{4(1 - k^2)}{k} \cdot k^2 = 4k - 4k^3$$

$$\sum_{k=1}^{n} A_k B_k = \sum_{k=1}^{n}(4k - 4k^3) = 4\sum_{k=1}^{n} k - 4\sum_{k=1}^{n} k^3 = 2n(n + 1) - n^2(n + 1)^2$$

$$= -(n - 1)n(n + 1)(n + 2)$$

$$\frac{(n - 2)!}{(n + 2)!} \sum_{k=1}^{n} A_k B_k = -\frac{(n - 2)!}{(n + 2)!}(n - 1)n(n + 1)(n + 2) = -\frac{(n - 1)n(n + 1)(n + 2)}{(n - 1)n(n + 1)(n + 2)} = \boxed{-1}$$

Exercise 6.8

$$B_k = 4k^2 \left(\frac{1 - k}{1 + k} - \frac{1 + k}{1 - k}\right)^{-1} + k = k^3$$

$$A_s = s - \frac{4}{s^2 + s}\left(1 + \sum_{k=2}^{s} k^3\right) = s - \frac{4}{s^2 + s}\left(1 + \frac{s^2(s + 1)^2}{4} - 1\right) = -s^2$$

$$\frac{1}{72n^2 + 18n + 1} \sum_{s=1}^{6n}(-s^2) = -\frac{n(12n + 1)(6n + 1)}{72n^2 + 18n + 1} = \boxed{-n}$$

Exercise 6.9

Using formula 6.12, we have:

$$\sum_{k=0}^{n-1} 2^k = 2^n - 1$$

$$\sum_{k=0}^{n-1} 4^k = \frac{4^n - 1}{3}$$

Therefore:

$$1 + \frac{3}{2^n - 2} \sum_{k=0}^{n-1}(4^k - 2^k) = \boxed{2^n}$$

Exercise 6.10

1.

$$\frac{2^{-x-1}(2x)!}{x!(2x-3)!!} + \frac{1}{2} = \frac{(2x)!}{x!2^{x+1}(2x-3)!!} + \frac{1}{2} = \frac{1}{2(2x-3)!!} \cdot \frac{(2x)!}{x!2^x} + \frac{1}{2}$$

Let's simplify $\frac{(2x)!}{x!2^x}$ separately:

$$\frac{(2x)!}{x!2^x} = \frac{1 \cdot 2 \cdot 3 \cdots (2x)}{x!(2 \cdot 2 \cdot 2 \cdots 2)} = \frac{1}{x!} \cdot \frac{2}{2} \cdot 3 \cdot \frac{4}{2} \cdot 5 \cdot \frac{6}{2} \cdot 7 \cdots \frac{2x-2}{2} \cdot (2x-1) \cdot \frac{2x}{2}$$

$$= \frac{x!}{x!} \cdot 3 \cdot 5 \cdot 7 \cdots (2x-1) = (2x-1)!!$$

Thus:

$$\frac{1}{2(2x-3)!!} \cdot \frac{(2x)!}{x!2^x} + \frac{1}{2} = \frac{(2x-1)!!}{2(2x-3)!!} + \frac{1}{2} = \frac{2x-1}{2} + \frac{1}{2} = x$$

2.

$$1 + \frac{(x-1)}{x} \sum_{k=1}^{n} \left(\frac{2^{-x-1}(2x)!}{x!(2x-3)!!} + \frac{1}{2} \right)^k = 1 + \frac{(x-1)}{x} \sum_{k=1}^{n} x^k$$

$$= 1 + \frac{(x-1)}{x} \left(x^n - 1 + \sum_{k=0}^{n-1} x^k \right) = 1 + \frac{(x-1)}{x} \left(x^n - 1 + \frac{x^n - 1}{x - 1} \right) = x^n$$

3.

$$\sqrt[n]{x^n} = \boxed{x}$$

Exercise 6.11

$$B_n = \sum_{k=-1}^{n} 2^{n+1} x^k = \frac{2^{n+1}}{x} + 2^{n+1} x^n + 2^{n+1} \sum_{k=0}^{n-1} x^k$$

$$= \frac{2^{n+1}}{x} + 2^{n+1} x^n + 2^{n+1} \cdot \frac{x^n - 1}{x - 1} = \frac{2^{n+1}(x^{n+2} - 1)}{(x-1)x}$$

$$C_n = \left(\sum_{s=3}^{n+1} 2x^s \right)^{-1} = \left(-2 - 2x - 2x^2 + 2x^n + 2x^{n+1} + 2 \sum_{s=0}^{n-1} x^s \right)^{-1}$$

$$= \left(-2 - 2x - 2x^2 + 2x^n + 2x^{n+1} + 2\frac{x^n - 1}{x - 1} \right)^{-1} = \frac{x-1}{2x(x^{n+1} - x^2)}$$

$$A_n B_n C_n = -\frac{x^2(x^{n+1} - x^2)}{x^{n+2} - 1} \cdot \frac{2^{n+1}(x^{n+2} - 1)}{(x-1)x} \cdot \frac{x-1}{2x(x^{n+1} - x^2)} = -2^n$$

$$\sum_{n=2}^{10}(-2^n) = -\sum_{n=2}^{10} 2^n = -(-1) - (-2) - \sum_{n=0}^{10} 2^n = -(-1) - (-2) - (2^{11} - 1) = \boxed{-2044}$$

Exercise 6.12

(a)

$$\sum_{i=0}^{n-1}(a + id) = a\sum_{i=0}^{n-1} 1 + d\sum_{i=0}^{n-1} i = an - dn + d\sum_{i=1}^{n} i$$

$$= an - dn + \frac{dn(n+1)}{2} = \boxed{an + \frac{dn(n-1)}{2}}$$

(b)

$$a_1 = a \qquad and \qquad a_n = a + d(n-1)$$

$$an + \frac{dn(n-1)}{2} = \frac{n(2a + dn - d)}{2} = \frac{n(a + a + d(n-1))}{2} = \boxed{\frac{n(a_1 + a_n)}{2}}$$

(c)

$$\frac{n(a_1 + a_n)}{2} = \frac{51(3 + 103)}{2} = \boxed{2703}$$

Or

$$\sum_{i=0}^{50}(3 + 2i) = 3 \cdot 51 + \frac{2 \cdot 51(51 - 1)}{2} = \boxed{2703}$$

(d)

$$\frac{n(a_1 + a_n)}{2} = \frac{23(108 + (-2))}{2} = \boxed{1219}$$

Or

$$\sum_{i=0}^{22}(108 + (-5)i) = 108 \cdot 23 + \frac{(-5) \cdot 23(23 - 1)}{2} = \boxed{1219}$$

(e)

$$\sum_{i=2}^{20}(3i - 100) = 100 + 97 + \sum_{i=0}^{20}(3i - 100) = 197 + (-100) \cdot 21 + \frac{3 \cdot 21(21 - 1)}{2} = \boxed{-1273}$$

Exercise 6.13

(a) Using formula 6.12:

$$\sum_{i=0}^{n-1} gr^i = g\sum_{i=0}^{n-1} r^i = \boxed{\frac{g(r^n - 1)}{r - 1}}$$

(b)

$$g_1 = g \qquad and \qquad g_n = gr^{n-1}$$

$$\frac{g(r^n - 1)}{r - 1} = \frac{gr^{n-1}r - g}{r - 1} = \boxed{\frac{g_n r - g_1}{r - 1}}$$

(c)

$$\frac{g_n r - g_1}{r - 1} = \frac{512 \cdot 2 - 2}{2 - 1} = \boxed{1022}$$

Or

$$\sum_{i=0}^{8} 2 \cdot 2^i = \frac{2(2^9 - 1)}{2 - 1} = 2^{10} - 2 = \boxed{1022}$$

(d)

$$\frac{g_n r - g_1}{r - 1} = \frac{384 \cdot 2 - 3}{2 - 1} = \boxed{765}$$

Or

$$\sum_{i=0}^{7} 3 \cdot 2^i = \frac{3(2^8 - 1)}{2 - 1} = \boxed{765}$$

(e)

$$\sum_{i=1}^{10} \frac{(-2)^i}{2} = \sum_{i=0}^{10} \frac{(-2)^i}{2} - \frac{1}{2} = \frac{0.5((-2)^{11} - 1)}{(-2) - 1} - \frac{1}{2} = \boxed{341}$$

Exercise 6.14

(a) Using Formulas 6.10 and 6.11:

$$\frac{n\sum_{i=1}^{n} i^2 - \sum_{i=1}^{n} i^3}{2n} = \frac{1}{2n}\left(n\frac{n(n+1)(2n+1)}{6} - \frac{n^2(n+1)^2}{4}\right) = \boxed{\frac{n(n^2 - 1)}{24}}$$

(b)

$$\frac{n(n^2 - 1)}{24} = \frac{24(24^2 - 1)}{24} = 24^2 - 1 = 576 - 1 = \boxed{575}$$

(c) The numbers form the arithmetic progression. The first element is -50 and the last element is 346. Using the formula for the sum of an arithmetic progression (see Exercise 6.12):

$$\mu = \frac{\sum_{i=1}^{n} x_i}{n} = \frac{n(-50 + 346)}{2 \cdot n} = \boxed{148}$$

(d)

$$\mu = \frac{\sum_{i=1}^{n} x_i}{n} = \frac{n(a_1 + a_n)}{2n} = \boxed{\frac{a_1 + a_n}{2}}$$

Thus, the mean of an arithmetic progression is the arithmetic mean of its first and last terms.

(e)

$$\frac{a_{i-1} + a_{i+1}}{2} = \frac{(a_1 + (i-2)d) + (a_1 + id)}{2} = \frac{2a_1 + 2id - 2d}{2}$$

$$= a_1 + id - d = a_1 + d(i-1) = a_i$$

Exercise 6.15

(a) The given population is an arithmetic progression. From Exercise 6.14(d), we know that:

$$\mu = \frac{a_1 + a_n}{2} = \frac{-10 + 12}{2} = 1$$

Now we are ready to find the variance. For this, we first evaluate the sum (using Formula 6.10):

$$\sum_{i=-10}^{12} (i - 1)^2 = \sum_{i=-11}^{11} i^2 = \sum_{i=-11}^{-1} i^2 + \sum_{i=1}^{11} i^2$$

$$= \sum_{i=1}^{11} (-i)^2 + \sum_{i=1}^{11} i^2 = 2 \sum_{i=1}^{11} i^2 = 2 \cdot \frac{11(11 + 1)(2 \cdot 11 + 1)}{6} = 1012$$

With this in mind:

$$\sigma^2 = \frac{\sum_{i=1}^{n} (x_i - \mu)^2}{n} = \frac{\sum_{i=-10}^{12} (i - 1)^2}{23} = \frac{1012}{23} = \boxed{44}$$

(b)

$$\sigma^2 = \frac{\sum_{i=1}^{n} (x_i - \mu)^2}{n}$$

$$= \frac{\sum_{i=1}^{n} (x_i^2 - 2x_i\mu + \mu^2)}{n} = \frac{\sum_{i=1}^{n} x_i^2}{n} - 2\mu \cdot \frac{\sum_{i=1}^{n} x_i}{n} + \mu^2 \cdot \frac{\sum_{i=1}^{n} 1}{n}$$

$$= \nu - 2\mu^2 + \mu^2 = \nu - \mu^2$$

Exercise 6.16

(a) Using 6.15:

$$\gamma = \sqrt[n]{\prod_{i=1}^{n} i^n} = \sqrt[n]{\left(\prod_{i=1}^{n} i\right)^n} = \prod_{i=1}^{n} i = \boxed{n!}$$

(b)

$$\gamma = \sqrt[n]{\prod_{i=1}^{n} 2^i} = \sqrt[n]{2^{1+2+3+\cdots+n}} = \sqrt[n]{2^{\frac{n(n+1)}{2}}} = \boxed{\sqrt{2^{n+1}}}$$

(c)

$$\gamma = \sqrt[n]{\prod_{i=0}^{n-1}(3 \cdot 2^i)} = \sqrt[n]{3^n \prod_{i=0}^{n-1} 2^i} = \sqrt[n]{3^n 2^{-n} \prod_{i=1}^{n} 2^i} = 3\sqrt[n]{2^{\frac{n(n+1)}{2}-n}} = \boxed{3\sqrt{2^{n-1}}}$$

(d)

$$\gamma = \sqrt[n]{\prod_{i=0}^{n-1}(g \cdot r^i)} = \sqrt[n]{g^n \prod_{i=0}^{n-1} r^i} = \sqrt[n]{g^n r^{-n} \prod_{i=1}^{n} r^i} = g\sqrt[n]{r^{\frac{n(n+1)}{2}-n}} = \boxed{g\sqrt{r^{n-1}}}$$

(e) We already know the geometric mean of a geometric progression from part (d). Let's rewrite it through the first and last terms of the progression:

$$\sqrt[n]{\prod_{i=1}^{n} g_i} = \sqrt[n]{\prod_{i=0}^{n-1}(g \cdot r^i)} = g\sqrt{r^{n-1}} = \sqrt{g}\sqrt{g}\sqrt{r^{n-1}} = \sqrt{ggr^{n-1}} = \boxed{\sqrt{g_1 g_n}}$$

Thus, the geometric mean of a geometric progression is the geometric mean of its first and last terms.

(f)

$$\sqrt{g_{i-1} g_{i+1}} = \sqrt{(g_1 r^{i-2})(g_1 r^i)} = \sqrt{g_1^2 r^{2i-2}} = g_1 r^{i-1} = g_i$$

(g) This is the geometric mean of the geometric progression. The last term is approximately equal to 4. Using the result we obtained in (e):

$$\gamma \approx \sqrt{1 \cdot 4} = \boxed{2}$$

Exercise 6.17

$$d^2 = \sum_{i=1}^{12}(i-1)^2 = \sum_{i=1}^{12} i^2 - 2\sum_{i=1}^{12} i + \sum_{i=1}^{12} 1 = \frac{12(12+1)(24+1)}{6} - 2\frac{12(12+1)}{2} + 12 = \boxed{506}$$

Exercise 6.18

(a)

$$P_5^7 = \frac{7!}{(7-5)!} = \frac{7!}{2!} = \frac{5040}{2} = \boxed{2520}$$

(b)

$$P_n^n = \frac{n!}{(n-n)!} = \boxed{n!}$$

(c)

$$5! = \boxed{120}$$

Exercise 6.19

(a)

$$C_5^7 = \frac{7!}{5!(7-5)!} = \frac{7!}{5!(7-5)!} = \frac{6 \cdot 7}{2} = \boxed{21}$$

(b)

$$C_{n-k}^n = \frac{n!}{(n-k)!(n-(n-k))!} = \frac{n!}{(n-k)!k!} = C_k^n$$

(c)

$$C_{k-1}^{n-1} + C_k^{n-1} = \frac{(n-1)!}{(k-1)!(n-k)!} + \frac{(n-1)!}{k!(n-k-1)!}$$

$$= \frac{k(n-1)!}{k!(n-k)!} + \frac{(n-k)(n-1)!}{k!(n-k)!} = \frac{(k+n-k)(n-1)!}{k!(n-k)!}$$

$$= \frac{n(n-1)!}{k!(n-k)!} = \frac{n!}{k!(n-k)!} = C_k^n$$

Exercise 6.20

(a)

$$(a+b)^2 = \sum_{k=0}^{2} C_k^2 a^{2-k} b^k$$

$$= C_0^2 a^{2-0} b^0 + C_1^2 a^{2-1} b^1 + C_2^2 a^{2-2} b^2 = 1 \cdot a^2 b^0 + 2 \cdot a^1 b^1 + 1 \cdot a^0 b^2$$

$$= \boxed{a^2 + 2ab + b^2}$$

$$(a+b)^3 = \sum_{k=0}^{3} C_k^3 a^{3-k} b^k$$

$$= C_0^3 a^{3-0} b^0 + C_1^3 a^{3-1} b^1 + C_2^3 a^{3-2} b^2 + C_3^3 a^{3-3} b^3$$

$$= 1 \cdot a^3 b^0 + 3 \cdot a^2 b^1 + 3 \cdot a^1 b^2 + 1 \cdot a^0 b^3$$

$$= \boxed{a^3 + 3a^2 b + 3ab^2 + b^3}$$

(b)

$$(a+b)^4 = \sum_{k=0}^{4} C_k^4 a^{4-k} b^k$$

$$= C_0^4 a^{4-0} b^0 + C_1^4 a^{4-1} b^1 + C_2^4 a^{4-2} b^2 + C_3^4 a^{4-3} b^3 + C_4^4 a^{4-4} b^4$$

$$= 1 \cdot a^4 b^0 + 4 \cdot a^3 b^1 + 6 \cdot a^2 b^2 + 4 \cdot a^1 b^3 + 1 \cdot a^0 b^4$$

$$= \boxed{a^4 + 4a^3 b + 6a^2 b^2 + 4ab^3 + b^4}$$

(c)

$$(a+b)^5 = \sum_{k=0}^{5} C_k^5 a^{5-k} b^k$$

$$= C_0^5 a^{5-0} b^0 + C_1^5 a^{5-1} b^1 + C_2^5 a^{5-2} b^2 + C_3^5 a^{5-3} b^3 + C_4^5 a^{5-4} b^4 + C_5^5 a^{5-5} b^5$$

$$= 1 \cdot a^5 b^0 + 5 \cdot a^4 b^1 + 10 \cdot a^3 b^2 + 10 \cdot a^2 b^3 + 5 \cdot a^1 b^4 + 1 \cdot a^0 b^5$$

$$= \boxed{a^5 + 5a^4 b + 10a^3 b^2 + 10a^2 b^3 + 5ab^4 + b^5}$$

(d)

$$(1+1)^n = \sum_{k=0}^{n} C_k^n 1^k 1^{n-k} = \sum_{k=0}^{n} C_k^n$$

(e)

$$
\begin{array}{c}
n = 0 \qquad\qquad 1 \\
n = 1 \qquad\quad 1 \quad 1 \\
n = 2 \qquad 1 \quad 2 \quad 1 \\
n = 3 \quad 1 \quad 3 \quad 3 \quad 1 \\
n = 4 \quad 1 \quad 4 \quad 6 \quad 4 \quad 1 \\
n = 5 \quad 1 \quad 5 \quad 10 \quad 10 \quad 5 \quad 1 \\
n = 6 \quad 1 \quad 6 \quad 15 \quad 20 \quad 15 \quad 6 \quad 1
\end{array}
$$

Thus:

$$(a+b)^6 = \boxed{a^6 + 6a^5 b + 15a^4 b^2 + 20a^3 b^3 + 15a^2 b^4 + 6ab^5 + b^6}$$

Index

www.ingramcontent.com/pod-product-compliance
Lightning Source LLC
Chambersburg PA
CBHW081809200326
41597CB00023B/4200